VARIATIONAL METHODS

for Boundary Value Problems for Systems of Elliptic Equations

BY

M. A. LAVRENT'EV

AUTHORIZED TRANSLATION
FROM THE RUSSIAN BY

J. R. M. RADOK

DOVER PUBLICATIONS, INC.
NEW YORK

Published in Canada by General Publishing Company, Ltd., 30 Lesmill Road, Don Mills, Toronto, Ontario.
Published in the United Kingdom by Constable and Company, Ltd., 10 Orange Street, London WC2H 7EG.

This Dover edition, first published in 1989, is an unabridged, slightly corrected republication of the work first published by P. Noordhoff, Ltd, Groningen, The Netherlands, 1963. The translator has prepared an Index specially for the Dover edition.

Manufactured in the United States of America
Dover Publications, Inc., 31 East 2nd Street, Mineola, N.Y. 11501

Library of Congress Cataloging-in-Publication Data

Lavrent'ev, Mikhail Alekseevich, 1900–
[Variat͡sionnyĭ metod v kraevykh zadachakh dli͡a sistem uravneniĭ ėllipticheskogo tipa. English]
Variational methods for boundary value problems for systems of elliptic equations / by M. A. Lavrent'ev ; authorized translation from the Russian by J. R. M. Radok.
p. cm.
Translation of: Variat͡sionnyĭ metod v kraevykh zadachakh dli͡a sistem uravneniĭ ėllipticheskogo tipa.
"This Dover edition, first published in 1989, is an unabridged, slightly corrected republication of the work first published by P. Noordhoff, Ltd, Groningen, The Netherlands, 1963. The translator has prepared an Index specially for the Dover edition."—CIP t.p. verso.
Includes bibliographical references.
ISBN 0-486-66170-9 (pbk.)
1. Calculus of variations. 2. Conformal mapping. 3. Boundary value problems. 4. Fluid dynamics. I. Radok, J. R. M. (Jens Rainer Maria) II. Title.
QA316.L36313 1989
515'.64—dc20 89-17054
CIP

FOREWORD

At the present time, the problems of the existence and stability of the solutions of different classes of boundary value problems of the equations of mathematical physics continue to be of great interest. In particular, great progress has occurred during the past ten years in the domain of linear problems, where the method of integral equations with Fredholm's well-known alternative made it possible to arrive at final solutions of all basic linear problems for equations of the elliptic type; this method also greatly assisted the advance of the study of Tricomi's problem for equations of the mixed type.

Beginning with the well-known studies of Villat, Levi-Cività and Nekrasov, we have a large group of investigations of classical non-linear problems of the mechanics of continuous media, of the problems of the flow around an arbitrary contour and of the wave motion of a heavy fluid. The majority of studies in this direction also employ non-linear integral equations with an application of the method of expansion with respect to a small parameter (Nekrasov, Kochin, etc.) or with an application of the methods of functional analysis, in particular, of the well-known fixed point theorem (Leray, Weinstein, Kravtchenko, etc.).

In this monograph, the author presents an approach to these problems which is different in principle. It is based on a number of geometric properties of conformal and quasi-conformal mappings and uses a general basic scheme for the solution of variational problems, first suggested by Hilbert and developed by Tonnelli.

The method lies on the boundary between the classical methods of analysis with its concrete estimates and approximate formulae and the methods of the theory of functions of a real variable with their special character and general theoretical quantitative aspects.

On the basis of these facts, I wish to emphasize beforehand, that in the writing of the monograph I have tended to organize it in such a way that it may be used by mathematicians as well as by theoreticians in mechanics, far removed from function theory. Probably, the method of the proof of the existence and uniqueness theorems will be of interest to mathematicians (Chapters I–III) as well as the general theory of quasi-conformal mappings; mathematicians may readily omit those sections which refer to approximate methods and studies of particular problems in hydrodynamics and wave theory.

Theoreticians in mechanics, far removed from function theory, can freely overlook the general scheme of the proof of the existence and uniqueness theorems and concentrate on the approximate formulae for conformal and quasi-conformal mappings which may turn out to be useful (if combined with the methods for estimating errors proposed in the monograph) for the solution of many concrete problems of the mechanics of continuous media.

I have not attempted to present the details of proofs and frequently have only given an outline of the general ideas. Likewise I have not aimed at the widest possible assumptions nor have I dwelt on the functional theoretical aspects of the problems.

In conclusion I wish to thank Professor B. V. Shabat for his editorial help and for his highly valued comments relating to this book.

Novosibirsk, 1960 M. A. Lavrent'ev

CONTENTS

INTRODUCTION

1. Variational principles

Let there be given a class E of admissible curves γ and in this class let there be defined a functional $I(\gamma)$ such that a definite numerical value of I corresponds to every curve γ of E. Let it be required to find γ_0 of E for which I assumes an extreme value. In order to prove the existence of such a curve γ_0, construct, first of all, a minimizing sequence

$$\gamma_1, \gamma_2, \cdots, \gamma_n, \cdots \tag{1}$$

of curves of E with the property that

$$\lim_{n\to\infty} I(\gamma_n) = \inf I(\gamma). \tag{2}$$

The existence theorem reduces now to the establishment of the following two properties of I:

1) one can select from the sequence γ_n a compact sequence such that the limiting curve also lies in E;
2) semi-continuity of I: if γ_0 is the limiting curve of γ_n, then $\lim_{n\to\infty} I(\gamma_n) \geq I(\gamma_0)$.

For the applications presented below we require a minor modification of this scheme. Again let there be given some class of admissible curves $E = \{\gamma\}$ and on this class non-negative continuous functionals $I(\gamma)$; it is required to prove that there exists in the class E a curve γ_0 for which

$$I(\gamma_0) = 0. \tag{3}$$

For the solution of this problem it is sufficient to establish the possibility of a subdivision in E of some compact subclass E_0 such that every curve γ of E_0 for which $I(\gamma) > 0$ can be varied in the following manner:

1°. the variation $\tilde{\gamma}$ belongs to E_0;

2°. for $\tilde{\gamma}$ the value of I would be smaller than for the initial curve, i.e.,

$$I(\tilde{\gamma}) < I(\gamma). \tag{4}$$

In fact, it follows from the compactness of E_0 that one can select from the minimizing sequence (1) a uniformly convergent sequence with the limit curve Γ; however, by virtue of the continuity of I and of (2), one has

$$I(\Gamma) = \inf I.$$

Hence we conclude that $I(\Gamma) = 0$. In fact, if we had $I(\Gamma) > 0$, then we could find, by 2°., a line $\tilde{\Gamma}$ for which

$$I(\tilde{\Gamma}) < I(\Gamma) = \inf I,$$

which is impossible.

2. Sufficient conditions

We will now present examples of classes of functionals for which the above scheme may be realized.

Select for the class $E_0 = \{\gamma\}$ the manifold of curves $y = y(x)$, $|x| < \infty$ satisfying the conditions [1]:

$$\begin{aligned} &1°.\ 0 \le y(x) \le 1; \\ &2°.\ |y'(x)| \le k_1; \\ &3°.\ |y'(x+h) - y'(x)| \le k_2|h|. \end{aligned} \tag{5}$$

Further, let there be given on E_0 functionals $I(\gamma, x)$ with the following properties:

1°. $I(\gamma, x)$ is continuous and satisfies a Hölder condition with respect to x:

$$|I(\gamma, x+h) - I(\gamma, x)| \le k|h|^\nu, \tag{6}$$

where k and ν depend only on the constants k_1 and k_2;

---

[1] It follows from 3°. that on any finite interval γ has almost everywhere a curvature k where $|y''(x)| \le k_2$.

2°. For a transition from γ to $\tilde{\gamma}$ at all points where $\tilde{y}(x) - y(x)$ attains a non-negative maximum (non-positive minimum), we have

$$I(\tilde{\gamma}, x) < I(\gamma, x) \qquad (I(\tilde{\gamma}, x) > I(\gamma, x)); \tag{7}$$

3°. $$\lim_{|x|\to\infty} I(\gamma, x) = 0; \tag{8}$$

4°. if γ contains a segment of a straight line with gradient $k_1(-k_1)$ or an arc of a parabola $y = ax^2/2 + bx + c$, $a = k_2$ $(a = -k_2)$, then the functional $I(\gamma, x)$ cannot attain a maximum at the right ends or a minimum at the left ends of the segment or arc, respectively; on replacing k_1 by $-k_1$ or k_2 by $-k_2$, respectively, the maxima and minima must change places;

5°. there exists a curve of E_0 for which $|I(\gamma, x)| < a$;

6°. if γ contains points on the straight lines $y = 0$ or $y = 1$, then at these points $I(\gamma, x) > a$ and $I(\gamma, x) < -a$, respectively, where $a > 0$.

THEOREM. *Under the conditions* 1°.–6°., *there exists in the class* E_0 *a curve* γ_0 *for which* $I(\gamma_0, x) \equiv 0$.

Proof. By virtue of the general variational principle formulated above, it is sufficient for the proof of this theorem to show that, whatever may be the curve γ_0 of E_0 for which $I(\gamma_0, x) \not\equiv 0$ or

$$\max |I(\gamma_0, x)| = \nu > 0,$$

one can always construct a new curve γ_0 in E_0 such that

$$\max |I(\gamma_0, x)| < \nu.$$

By virtue of 2°., one may assume beforehand for this purpose that

$$\max I(\gamma_0, x) = -\min I(\gamma_0, x) = \nu. \tag{9}$$

In fact, it follows from 5°. that $\nu < a$; therefore γ_0 has no common points with the straight lines $y = 0$ and $y = 1$, because this would contradict 6°. Using this fact one may select for $\tilde{\gamma}_0$ the curve γ_0 deflected in the direction of the y-axis by a sufficiently small amount; then, by 2°., one obtains for $\tilde{\gamma}_0$ that $\max |I(\tilde{\gamma}_0, x)| < \nu$.

Thus, let $\gamma_0 : y = y_0(x)$ be an arbitrary curve of E_0. Denote by E_+ and E_-, respectively, the sets of points where $I(\gamma_0, x)$ attains maxima and minima; we have for E_+ and E_- the condition (9). Both sets are closed and, by virtue of the uniform convergence of $I(\gamma, x)$, they lie at a finite distance from each other. Consequently, E_+ and E_- can contain the system of segments δ_+, δ_- with the properties:

1) between any two δ_+ lies a δ_- and between two δ_- lies a δ_+;
2) between the δ_+ and δ_- lie segments g_+ of length larger than ρ, where ρ does not depend on δ, and between the δ_- and δ_+ lie segments g_- the lengths of which are also larger than ρ.

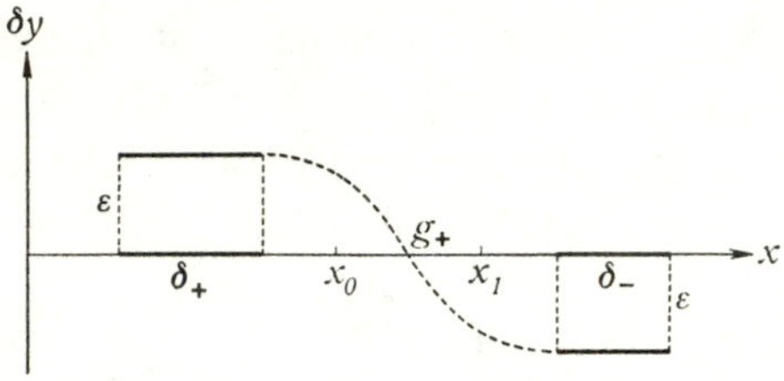

Fig. 1

Noting these facts we proceed to the construction of the unknown variation

$$\delta y = \tilde{y}(x) - y_0(x).$$

First of all, we determine δy in the intervals δ_+ and δ_- by letting

$$\delta y = +\,\varepsilon \text{ in all } \delta_+,$$
$$\delta y = -\,\varepsilon \text{ in all } \delta_-.$$

By 2°., if we predetermine δy in an arbitrary manner in the intervals g under the condition $|\delta y| < \varepsilon$, then we will have for the curve γ at the points of E_+ and E_-

$$\max |I(\tilde{\gamma}, x)| < \nu. \tag{10}$$

In order to prove the theorem, we must predetermine δy in the following manner:

a) the line $\tilde{\gamma}$ lies in E_0;
b) the inequality (10) is fulfilled for all x.

We will begin with the simplest case. Assume that all intervals g_+ and g_- contain segments $x_0 \leq x \leq x_1$ in which simultaneously

$$|y'(x)| \leq k_1' < k_1,$$
$$\lim_{h\to 0}\left|\frac{y'(x+h)-y'(x)}{h}\right| \leq k_2' < k_2. \tag{11}$$

Then we select in the capacity of δy in each segment g_+

$$\delta y = \varepsilon \cos \pi \frac{x-x_0}{x_1-x_0} \quad \text{for } x_0 \leq x \leq x_1,$$

$\delta y = \varepsilon$ at the points of g_+ which lie to the left of x_0,

$\delta y = -\varepsilon$ at the points of g_+ which lie to the right of x_1.

In every interval g_-, let

$$\delta y = -\varepsilon \cos \pi \frac{x-x_0}{x_1-x_0} \quad \text{for } x_0 \leq x \leq x_1,$$

$\delta y = \mp \varepsilon$ to the left of x_0 and to the right of x_1, respectively.

By 3°., the number of segments δ_+ and δ_- is finite; on the left side of the most left segment, let $\delta y = \pm \varepsilon$, to the right of the most right segment, $\delta y = \pm \varepsilon$, the signs being selected in dependence on whether the corresponding intervals belong to δ_+ or δ_-.

It is readily shown that for sufficiently small ε the varied curve will satisfy all the conditions below.

Now we will proceed to the general case. Let a and b be the ends of the interval g_+. Predetermine γ in the neighbourhood of the point b. It follows from 4°. that in any neighbourhood of the point b there is a point b' fulfilling the following conditions:

1) at b', and consequently also in some neighbourhood of this point, we have $y'(x) > -k_1$;

2) the point b' is a limit point of the set of points for which $y''(x)$ exists and is less than k_2.

With these facts in mind construct the parabola

$$y = \tfrac{1}{2}k_2x^2 + bx + c = Y(x)$$

tangential to γ_0 at a point with abscissa b'. By 1) and 2), we will have for any point b'' sufficiently close to the left of b' that

$$|Y'(b'')| < k_1, \quad Y(b'') > y(b'').$$

We now construct the parabola

$$y = -\tfrac{1}{2}k_2x^2 + b'x + c' = Y_1(x)$$

which has for $x = b''$ with the parabola Y a common tangent and ordinate. Using 2) we find now a point b''' for which

$$y'(b''') = Y_1'(b'''),$$

where

$$Y(b''') - y(b''') = \varepsilon_+ > 0.$$

We note still that for $b'' \to b'$ the point b''' will likewise tend to b', and ε_+ will tend to zero, so that for sufficiently small ε a value of b'' can be found for which $\varepsilon_+ = \varepsilon$.

In a completely analogous manner one can construct corresponding parabolae to the right of the point a for the intervals $a' < x < a''$ and $a'' < x < a'''$, respectively. Now design for g_+ the varied curve $\tilde{\gamma} : y = \tilde{y}(x)$ in the following manner:

$$\begin{aligned}
\tilde{y}(x) &= y(x) + \varepsilon, & a &< x < a', \\
\tilde{y}(x) &= Y_2(x) + \varepsilon, & a' &< x < a'', \\
\tilde{y}(x) &= Y_3(x) + \varepsilon, & a'' &< x < a''', \\
\tilde{y}(x) &= y(x), & a''' &< x < b''', \\
\tilde{y}(x) &= Y_1(x) - \varepsilon, & b''' &< x < b'', \\
\tilde{y}(x) &= Y_2(x) - \varepsilon, & b'' &< x < b', \\
\tilde{y}(x) &= y(x) - \varepsilon, & b' &< x < b.
\end{aligned}$$

It follows directly from this arrangement that the curve $\tilde{\gamma} : y = \tilde{y}(x)$ belongs in g_+ to the class E_0 [1]. On the other hand, by 3°., the number of segments δ_+ and δ_- is bounded and for sufficiently small ε the construction is possible for all finite g; for $g_{\pm\infty}$ ($g_{-\infty}$ extending to the left of the most left δ and $g_{+\infty}$ to

[1] The conditions 4°. and 5°. together ensure that γ belongs to the strip $0 \leq y \leq 1$.

the right of the most right δ), we let

$$\tilde{y}(x) = y_0(x) \pm \varepsilon,$$

where the $\pm$ signs must be selected in correspondence with the fact whether the interval belongs to δ_+ or δ_-.

The entire curve $\tilde{\gamma}$ belongs to E_0. We still have to show that the inequality (10) is true for this curve. This inequality will be satisfied in all δ_+ and δ_-; by virtue of the continuity of $I(\gamma, x)$ with respect to x, it will also be fulfilled (for sufficiently small ε and a', b' sufficiently close to a, b) in the intervals aa', bb'. We still have to study the intervals $a'b'$ and $g_{\pm\infty}$; the fulfillment of (10) in these regions is ensured by 3°. and the continuity of I with respect to γ.

3. Generalizations

Naturally, the class of admissible curves can be changed in dependence on the boundary value problem. For example, in the problem of stream line flow around an arc Γ (to be considered in Chapter III) one finds it to be convenient to select as admissible curves the set of curves $y = y(x)$ with the properties:

1°. $y(x)$ is determined for $x \geq 0$, $y(0) = 1$, $y(x) > 0$;

2°. $|y'(x)| \leq k_1$, $|y'(x + h) - y'(x)| \leq k_2|h|^\alpha$;

3°. the curvature K of the curve γ satisfies the inequality $K \leq k_3/x$;

4°. $0 \leq y(x) \leq 1$.

In correspondence with a change of the class of admissible curves, one has also to change the conditions imposed on the functional $I(\gamma, x)$ so that increasing $\max |I|$ for variation of γ will be possible in the class of admissible curves under consideration.

The choice of the class of admissible curves is determined, first of all, by the fact that this class must contain the solution sought; in particular, the conditions 2°. and 3°. were designed such that particular solutions of the problem of jets satisfy these conditions. In the problem of waves in a viscous fluid, one must select as class of admissible curves the set of curves $y = y(x)$ which exhibit the periodicity $y(x + \omega) = y(x)$.

CHAPTER I

VARIATIONAL PRINCIPLES OF THE THEORY OF CONFORMAL MAPPING

1.1. The principles of Lindelöf and Montel

Let there be given in the plane of the complex variable z two simply connected regions D and $\tilde{D}$, bounded by curves Γ and $\tilde{\Gamma}$, and let $w = f(z)$ and $w = \tilde{f}(z)$ be functions which map the regions D and $\tilde{D}$ on to one of the standard regions: circle, half-plane or strip. As supplementary conditions which determine the mapping uniquely let us require that for $\tilde{\Gamma} \equiv \Gamma$ we have $\tilde{f} \equiv f$.

At the basis of the application of variational methods to the theory of functions of a complex variable or to problems of mechanics which can be solved by the methods of this theory we have the following problem.

Assuming $w = f(z)$ to be known and $\tilde{\Gamma}$ to be infinitely close to Γ, find the variation of $f(z)$, i.e., the principal linear part of the variation of $f(z)$ as $\Gamma \to \tilde{\Gamma}$.

We will consider the cases of mappings on to the circle, half-plane and strip separately for the corresponding, most frequently encountered normalisations.

1.1.1. The case of the circle. Let $D = D(\Gamma)$ denote the simply connected region bounded by the line Γ. Select in D some fixed point z_0 and map D conformally on to the unit circle $|w| < 1$ so that the point z_0 corresponds to the point $w = 0$:

$$w = f(z, \Gamma), \; f(z_0, \Gamma) = 0. \tag{1.1}$$

There will be infinitely many functions having this property, but all of them will differ by factors $e^{i\theta}$, where θ is an arbitrary, real number. The multiplier $e^{i\theta}$ will not play any role in what follows and we will understand by f any function of this class.

We will denote the closed curve corresponding to the transformation (1.1) to the circle $|w| = r < 1$ by γ_r; on replacing Γ by $\tilde{\Gamma}$, the corresponding function f and curve γ_r will be denoted by $\tilde{f}$ and $\tilde{\gamma}_r$, respectively.

Consider the polar coordinate system r, φ with the pole at the point z_0 and assume that in this system of coordinates the radii r and $\tilde{r}$ of the points of Γ and $\tilde{\Gamma}$ are single-valued functions of φ (i.e., the regions $D(\Gamma)$ and $D(\tilde{\Gamma})$ form rings around z_0). We will call the points $z_1 = r_1 e^{i\varphi_1}$ of the contour Γ, at which $\lambda = r(\varphi)/\tilde{r}(\varphi)$ attains a maximum and $z_2 = r_2 e^{i\varphi_2}$, where λ attains a minimum points of largest deformation of Γ (with respect to z_0); the corresponding numbers λ_1 and λ_2 will be called upper and lower bounds of deformation.

We will now formulate

THEOREM 1.1 (Lindelöf's principle). *If the region $D(\tilde{\Gamma})$ is contained in $D(\Gamma)$, the following results hold:*

1°. *for any $r < 1$, the region $D(\tilde{\gamma}_r)$ is contained in $D(\gamma_r)$, where γ_r and $\tilde{\gamma}_r$ can only coincide when $\tilde{\Gamma}$ and Γ coincide;*

2°. *at the point z_0*

$$|f'(z_0, \tilde{\Gamma})| \geq |f'(z_0, \Gamma)|, \tag{1.2}$$

where equality occurs only for $\tilde{\Gamma} \equiv \Gamma$;

3°. *if Γ and $\tilde{\Gamma}$ have a point z_1 in common and at this point f' and $\tilde{f}'$ exist, then*

$$|f'(z_1, \tilde{\Gamma})| \leq |f'(z_1, \Gamma)|; \tag{1.3}$$

4°. *if Γ and $\tilde{\Gamma}$ are annular with respect to z_0, then at the point where the upper bound of the largest deformation is attained* ($\lambda > 1$)

$$|f'(\tilde{z}_2, \tilde{\Gamma})| \geq \lambda |f'(z_2, \Gamma)|. \tag{1.4}$$

In other words, if we consider regions containing the fixed point z_0 and their conformal mappings on to the unit circle for which z_0 goes over into the centre of the circle, then in the case of indentation of the boundaries:

1°. all equipotential lines move;

2°. the extension at the point z_0 increases;

3°. at points of the boundaries which are not deformed the extension decreases;

4°. at points of maximum deformation the extension increases.

Proof. In the region $D(\Gamma)$, consider the harmonic function $p(x, y) = \log |f(z)/z|$; by assumption, this function is regular in D and assumes on Γ the value $\log 1/|z|$. If $D(\tilde{\Gamma})$ belongs to $D(\Gamma)$, one has on $\tilde{\Gamma}$

$$\log |f(z)| \leq 0, \ \log |\tilde{f}(z)| = 0;$$

consequently, by the maximum principle for harmonic functions, one has everywhere inside $D(\tilde{\Gamma})$ (if $\tilde{\Gamma} \not\equiv \Gamma$)

$$\log |f(z)| < \log |\tilde{f}(z)|. \tag{1.5}$$

On the other hand,

$$\log |f(z)| = |\tilde{f}(z)| = \log r$$

render the equations of the equipotential lines γ and $\tilde{\gamma}_r$; taking into consideration (1.5), we conclude therefore that $D(\tilde{\gamma}_r)$ is contained in $D(\gamma_r)$ (for $\tilde{\Gamma} \not\equiv \Gamma$).

The results 2°. and 3°. of Theorem 1.1 are obtained as limiting cases for $r \to 0$ and $r \to 1$, respectively; in order to derive 4°., it is sufficient to increase, in addition, the region $D(\tilde{\Gamma})$ λ times and to use 3°.

It is not difficult to derive from Lindelöf's principle another principle, which is also useful for applications:

THEOREM 1.2 (Montel's principle). *Let the regions $D(\Gamma)$ and $D(\tilde{\Gamma})$ contain the point z_0 and let Γ consist of arcs γ_1 and γ_2 such that γ_1 lies in $D(\tilde{\Gamma})$ and γ_2 outside $D(\tilde{\Gamma})$; in addition, let $\tilde{\Gamma}$ be obtained from Γ by replacement of γ_1 by the curve $\tilde{\gamma}_1$ lying outside $D(\Gamma)$, and of γ_2 by $\tilde{\gamma}_2$ lying in $D(\Gamma)$* (Fig. 2). *Further, let the transformations*

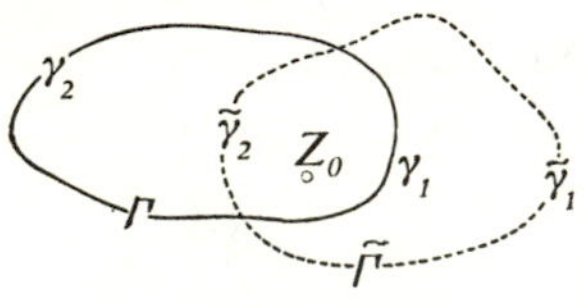

Fig. 2

$$w = f(z, \Gamma), \ f(z_0, \Gamma) = 0,$$
$$w = f(z, \tilde{\Gamma}), \ f(z_0, \tilde{\Gamma}) = 0$$

establish correspondence between the arcs γ_1 *and* $\tilde{\gamma}_1$ *and the arcs* φ_1 *and* $\tilde{\varphi}_1$ *of the unit circle. Then*

$$\varphi_1 \geq \tilde{\varphi}_1, \tag{1.6}$$

where the equality sign is only possible for $\tilde{\Gamma} \equiv \Gamma$.

In the case when the arcs γ_1, $\tilde{\gamma}_1$ are sufficiently smooth so that f' and $\tilde{f}'$ exist on the boundary, Theorem 1.2 follows from 3°. of Theorem 1.1; in the general case, the theorem may be derived either by considering an arbitrary region as a limiting case of smooth regions or by reducing the problem directly to the case of distortion of the unit circle.

1.1.2. Mapping on to a strip. Let Γ_0 and Γ be two smooth arcs the only common points of which are their ends a_1 and a_2; we will not exclude cases when one or both ends coincide with the point at infinity $z = \infty$. We will denote by $D(\Gamma_0, \Gamma)$ the region bounded by Γ_0 and Γ, and by

$$w = f(z, \Gamma_0, \Gamma), \; f(a_{1,2}, \Gamma_0, \Gamma) = \pm \infty \tag{1.7}$$

the function mapping the region $D(\Gamma_0, \Gamma)$ conformally on to the strip $0 < v < 1$. Under these conditions the function f will be determined apart from a real constant which is unimportant for the following considerations. Now let γ_v be the curve which becomes under the transformation (1.7) the straight line $v = \text{const}$.

For the mapping function $f(z, \Gamma_0, \Gamma)$ one has a principle which is analogous to Lindelöf's principle:

THEOREM 1.3. *If the region* $D(\tilde{\Gamma}_0, \Gamma)$ *is contained in* $D(\Gamma_0, \Gamma)$, *then*

1°. *for any* v, $0 < v < 1$, *the region* $D(\tilde{\gamma}_v, \Gamma)$ *is contained inside* $D(\gamma_v, \Gamma)$;

2°. *at any point* z_0 *of the curve* Γ

$$|f'(z_0, \tilde{\Gamma}_0, \Gamma)| \geq |f'(z_0, \Gamma_0, \Gamma)|;$$

3°. *if the curves* $\tilde{\Gamma}_0$ *and* Γ_0 *have a common point* z_1, *then one has at this point*

$$|f'(z_1, \tilde{\Gamma}_0, \Gamma)| \leq |f'(z_1, \Gamma_0, \Gamma)|$$

(*in all three cases the equality sign occurs only for* $\tilde{\Gamma}_0 \equiv \Gamma_0$);

4°. *in addition, let it be known that all the curves* Γ_0, $\tilde{\Gamma}_0$, Γ *are represented by single-valued functions* $y = f_0(x)$, $y = \tilde{f}_0(x)$, $y = f(x)$; *then we have at the point* $z_2 = x_2 + iy_0(x_2)$, *where* $f_0(x) - \tilde{f}_0(x)$ *attains a maximum, and at the corresponding point* $\tilde{z}_2 = x_2 + i\tilde{y}_0(x_2)$

$$|f'(\tilde{z}_2, \tilde{\Gamma}_0, \Gamma)| > |f'(z_2, \Gamma_0, \Gamma)|.$$

Proof. Consider the harmonic function

$$V = \operatorname{Im} f(z, \Gamma_0, \Gamma).$$

It is valid in $D(\Gamma_0, \Gamma)$ and assumes on Γ_0 and Γ values which are equal to 0 and 1, respectively. By virtue of the maximum principle, the function V is positive everywhere inside $D(\Gamma_0, \Gamma)$ and $0 < V < 1$; hence everywhere inside $D(\tilde{\Gamma}_0, \Gamma)$

$$V < \tilde{V}. \tag{1.8}$$

On the other hand, one has everywhere along the lines γ_v and $\tilde{\gamma}_v$

$$V = v, \quad \tilde{V} = v. \tag{1.9}$$

Comparing (1.8) and (1.9), we obtain 1°. In order to derive 2°., it is sufficient to note that on Γ

$$|f'(z, \Gamma_0, \Gamma)| = \frac{\partial V}{\partial n},$$

where $\partial/\partial n$ is the derivative in the direction of the normal to Γ.

1.2. Mechanical interpretation

The above principle has a very obvious mechanical interpretation. The function $w = f(z, \Gamma_0, \Gamma)$ may be considered as the complex potential of a velocity field of the steady motion of an ideal incompressible fluid. The motion is assumed to be two-dimensional, the liquid being contained between curvilinear walls Γ_0 and Γ. The width of the strip of the w-plane on to which the function f maps the region of flow is determined by the amount of flow of the fluid, i.e., by the quantity of fluid passing through the section in unit time; in our case this amount is equal to unity. The curves γ_v are the stream lines of the flow where the amount of flow between Γ_0 and γ_v is equal to v; in future, we will say that γ_v is the stream line corresponding to the amount v.

In correspondence with established hydrodynamic terminology the variational principle may be given the following formulation:

If the wall Γ_0, bounding the flow of a liquid, is pressed against the flow, then:

1°. all stream lines are brought closer to the other wall;

2°. the velocity of the flow along the untouched wall is everywhere increased;

3°. the velocity of the flow near the undeformed part Γ_0 is decreased;

4°. on Γ_0, at the location of the greatest contraction, the flow velocity is increased.

1.3. Quantitative estimates

The variational principle studied above permits us to estimate the sign of the variation as a function of the character of the variation of the boundary of the mapped region. In this way we can construct upper bounds for different functionals. We will begin with applications of the principle to the theory of approximate mappings, to the construction of variations and estimates of the remainder term.

Thus, assume that we know the function $w = f(z, \Gamma_0, \Gamma)$ and let it be required to construct its variation for the transition from $D(\Gamma_0, \Gamma)$ to $D(\Gamma_0, \tilde{\Gamma})$. Without restricting generality we can assume that $\tilde{\Gamma}$ is contained inside $D(\Gamma_0, \Gamma)$; in addition, we will consider only regions $D(\Gamma_0, \Gamma)$ for which $f'(z, \Gamma_0, \Gamma)$ exists and is continuous in D, including its boundary (the conditions which must be satisfied by Γ_0 and Γ for this case will be formulated in the following chapter).

In the case of the transformation $\omega = f(z, \Gamma_0, \Gamma)$, $\omega = \xi + i\eta$, the curve $\tilde{\Gamma}$ goes over into the curve L which lies close to the straight line $\eta = 1$; let the function $w = F(\omega, L)$, $F(\pm\infty, L) = \pm\infty$ bring about the mapping of the region $D(L)$, bounded by the straight line $\eta = 0$ and L, on to the strip $0 < v < 1$; then

$$f(z, \Gamma_0, \tilde{\Gamma}) = F[f(z, \Gamma_0, \Gamma), L]. \tag{1.10}$$

The function F maps the region "close" to the strip $0 < \eta < 1$ on to the strip $0 < v < 1$; hence

$$F = \omega + \varphi(\omega, L), \tag{1.11}$$

where φ is a "small" function. Substituting in (1.10) the expression for F from (1.11), we find

$$f(z, \Gamma_0, \tilde{\Gamma}) = f(z, \Gamma_0, \Gamma) + \varphi(f, L), \delta f(z, \Gamma_0, \Gamma) = \varphi(f, L). \tag{1.12}$$

Our problem has been reduced to the determination of the linear part of the function φ. Differentiating (1.12), one can obtain the variation for f'

$$f'(z, \Gamma_0, \tilde{\Gamma}) = \{1 + \varphi'(\omega, L)\}f'(z, \Gamma_0, \Gamma)$$

or

$$\delta \log f'(z, \Gamma_0, \Gamma) = \varphi'(\omega, L). \tag{1.13}$$

We will now compute the imaginary part of F to be denoted by V. On the curve L: $\eta = 1 - \lambda(\xi)$, the function V must assume the value unity, on the axis $\eta = 0$, the value zero. Let $\tilde{V}$ denote the harmonic function which assumes for $\eta = 0$ the value zero, and on the straight line $\eta = 1$ the value $1 + \lambda(\xi)$; then $\tilde{V}$ has on the line L the value

$$\tilde{V} = [1 + \lambda(\xi)] - \frac{\partial \tilde{V}}{\partial \eta} \lambda(\xi);$$

however,

$$\frac{\partial \tilde{V}}{\partial \eta} \approx 1 + \varepsilon,$$

where ε is small for sufficiently small and smooth functions $\lambda(\xi)$. Assume that $V = \tilde{V}$. By a well-known formula of the theory of conformal mapping (M. A. Lavrent'ev and B. V.

Shabat, 1958, p. 213), one has

$$V = \eta + \operatorname{Im} \tfrac{1}{2} \int_{-\infty}^{+\infty} \lambda(t) \tanh \frac{\pi(\omega - t)}{2}\, dt, \tag{1.14}$$

and hence

$$F(\omega) \approx \omega + \tfrac{1}{2} \int_{-\infty}^{+\infty} \lambda(t) \tanh \frac{\pi(\omega - t)}{2}\, dt.$$

In particular, for $\eta = 0$,

$$F'(\xi) \approx 1 + \frac{\pi}{4} \int_{-\infty}^{+\infty} \frac{\lambda(t)\, dt}{\cosh^2 \dfrac{\pi(\xi - t)}{2}}. \tag{1.15}$$

Thus, we obtain for the required variations formally the formulae: inside the region:

$$\delta f = \tfrac{1}{2} \int_{-\infty}^{+\infty} \lambda(t) \tanh \frac{\pi(\omega - t)}{2}\, dt, \tag{1.16}$$

for points where $\eta = 0$:

$$\delta \log |f'| = \frac{\pi}{4} \int_{-\infty}^{+\infty} \frac{\lambda(t)\, dt}{\cosh^2 \dfrac{\pi(\xi - t)}{2}}, \tag{1.17}$$

for points of L:

$$\delta \log |f'| = -\frac{\pi}{4} \int_{-\infty}^{+\infty} \frac{\lambda(t) - \lambda(\xi)}{\sinh^2 \dfrac{\pi(\xi - t)}{2}}\, dt. \tag{1.18}$$

Next, consider the problems of applicability and of estimates of remainder terms in the case when the increments of our functionals are replaced by their variations. If the integrals (1.14) and (1.16) are interpreted as Cauchy principal values, it will be sufficient for the purpose of making these integrals meaningful if the function $\lambda(\xi)$ satisfies a Hölder condition

$$|\lambda(\xi + h) - \lambda(\xi)| < k|h|^\alpha,\ k > 0,\ 0 > \alpha \leq 1;$$

for (1.18) to exist, the same condition must be satisfied by $\lambda'(\xi)$.

By virtue of the maximum principle, the accuracy in the definition of V will be completely determined by the accuracy of its boundary values, i.e., by the deviation of $\tilde{V}$ on L from unity. In order to derive the estimate desired it is sufficient to solve the following problem: a harmonic function $P(\xi, \eta)$ is equal to zero for $\eta = 0$, and equal to $\lambda(\xi)$ on the straight line $\eta = 1$; it is required to obtain an estimate of max $P[\xi, 1-\lambda(\xi)]$. By the same maximum principle, we obtain an estimate of the upper bound, if we replace $\lambda(\xi)$ everywhere by its maximum value $\max \lambda(\xi) = \nu$; but then we obtain $P = \nu\eta$, whence

$$|\delta V - [\tilde{V} - V]| < \nu^2. \tag{1.19}$$

For the function W, conjugate to P, we find everywhere

$$|\delta W - [\tilde{W} - W]| < \nu^2|\xi - \xi_0|, \tag{1.20}$$

where the point ξ_0 is determined by the correspondence of two given boundary points of Γ_0 and the axis u, selected beforehand.

The estimate (1.19) is preserved in the case of a variation of the normal derivative $\partial P/\partial n$ along the undeformed part of the boundary, i.e., for the points of Γ_0 and the corresponding straight line $u = 0$.

We will seek an estimate of the accuracy of (1.18). In order to find such an estimate, we must impose additional limitations on L: we will assume that $\lambda(x)$ has a small derivative and that this derivative satisfies a Hölder condition

$$|\lambda'(x + h) - \lambda'(x)| < \nu_2|h|^\alpha,\ 0 < \alpha \leq 1.$$

For these supplementary assumptions, by appealing to the maximum principle, we can readily construct an upper bound for $I = \left|\delta \frac{\partial w}{\partial n} - \left[\frac{\partial \bar{w}}{\partial n} - \frac{\partial w}{\partial n}\right]\right|$. In the capacity of the required majorant, one can select the normal derivative of the harmonic function P which assumes on $\eta = 0$ the value zero and on $\eta = 1$ the value

$$P = \nu_0 - \frac{\nu_2}{1+\alpha} |\xi|^{1+\alpha}$$

for $\nu_2 |\xi|^\alpha < \nu_1$,

$$P \text{ is linear, } P' = \pm \nu_1,$$

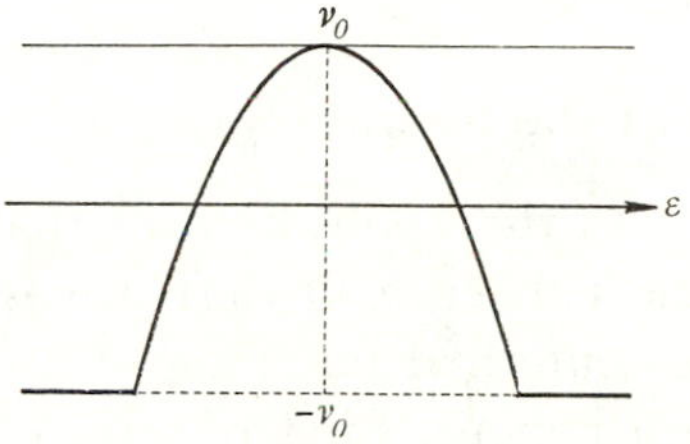

Fig. 3

up to the value $P = -\nu_0$,

$$P = -\nu_0 \text{ at all other points}$$

(fig. 3). Then, by (1.18), one has for the unknown upper bound the expression

$$I = \frac{\pi}{4} \int_{-\infty}^{+\infty} \frac{P\,dt}{\sinh^2 (t/2)}.$$

We have obtained estimates of the remainder terms in the case of replacement of the total increments of the functionals by their variations as functions of the properties of the curve L or of its mapping function $\eta = 1 - \lambda(\xi)$. It will be of some interest to compare these properties with those of the principal curves Γ_0, Γ, $\tilde{\Gamma}$: what limitations must be imposed on the curves Γ_0, Γ, $\tilde{\Gamma}$ so that the curve L will satisfy the corresponding conditions?

CHAPTER II

BEHAVIOUR OF A CONFORMAL TRANSFORMATION ON THE BOUNDARY

2.1. Derivatives at the boundary

We proceed now to the application of the basic principle to the problem of the existence of derivatives at the boundary in the case of the conformal mapping of a given region on to one of the standard regions. First of all, we note that it is indifferent for our problem under what conditions the mapping can be realized as well as whether the transformed region is a circle or a strip. Thus, let there be given a region $D(\Gamma)$, where Γ is a simple closed curve, and let

$$w = f(z, \Gamma), \; f(z_0, \Gamma) = 0 \tag{2.1}$$

be the function mapping D on the unit circle $|w| < 1$. Our problem is now as follows: what are the minimal conditions to be satisfied by Γ in order that

1) for $z \to z_1$, where z_1 is a point of Γ, $|f'(z)|$ will be bounded?

2) $|f'(z)|$, considered as a function of the points of Γ, will be continuous and satisfy a Hölder condition? We present here the fundamental theorems relating to this problem.

THEOREM 2.1. *If Γ is such that one may pass through its point z_1 two arcs γ_1 and γ_2 for each of which the angle formed with the tangent, considered as a function of the arc length, satisfies a Hölder condition in such a manner that one of them lies inside D, the other outside D* (fig. 4), *then for $z \to z_1$ along any path which is not tangential to Γ the upper bound of $|\log f'(z)|$ will be bounded.*

If the boundary Γ contains the arc γ which forms with the tangent an angle satisfying, as a function of the arc length, a Hölder condition, then at each point of γ the derivative of $f(z)$ exists and satisfies *along the arc the Hölder condition*

$$|f'(z + \Delta z) - f'(z)| < k|\Delta z|^\alpha. \tag{2.2}$$

Proof. First of all, we note that both statements of the theorem bear a local character and do not depend on what point corresponds to the point $w = 0$.

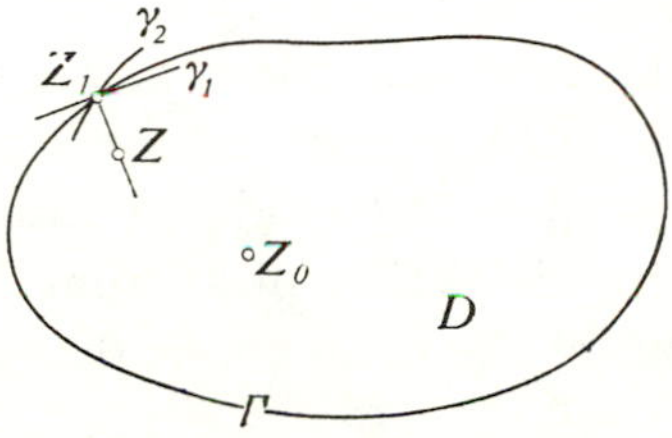

Fig. 4

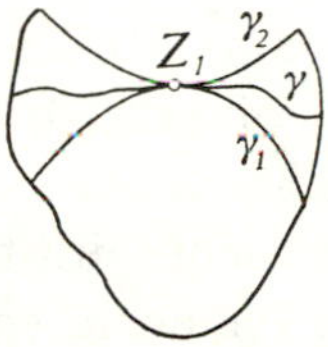

Fig. 5

The proof of the first part of the theorem will be given first. Construct three regions: $\Delta_1, \Delta, \Delta_2$ (fig. 5) containing the segments $\gamma_1, \gamma \subset \Gamma, \gamma_2$ of the boundary in such a manner that $\Delta_1 \subset \Delta \subset \Delta_2$. In addition, we can assume without reducing generality that all these regions contain the point z_0. By virtue of the statement 1°. of Lindelöf's principle, we have for $z \to z_1$ along a non-tangential path

$$\left| \frac{f_1(z) - f_1(z_1)}{z - z_1} \right| \leq \left| \frac{f(z) - f(z_1)}{z - z_1} \right| \leq \left| \frac{f_2(z) - f_2(z_1)}{z - z_1} \right|,$$

where f_1, f, f_2 are functions mapping the regions Δ_1, Δ, Δ_2 on to the unit circle. On the other hand, one can take for γ_1 and γ_2 the actual curves $y = |x|^\alpha$, $y = -|x|^\alpha$, so that f_1 and f_2 can be expressed in terms of elementary functions with continuous derivatives f_1' and f_2' up to the boundary. Hence follows the result of the first part of the theorem.

One of the following methods may be applied for the proof of the second part of the theorem:

Displacement method: In essence, this method involves the following. Let z_1 and z_2 be two adjacent points of the arc γ, $\gamma \subset \Gamma$.

Translate the region $D(\Gamma)$ by the vector $z_1 - z_2$ and rotate the region about the point z_1 in such a manner that the boundary $\tilde{\Gamma}$ of the rotated region D touches Γ at the point z_1; one has, obviously,

$$\Delta|f'| = |f'(z_2, \Gamma)| - |f'(z_1, \Gamma)| = |f'(z_1, \tilde{\Gamma})| - |f'(z_1, \Gamma)|.$$

The problem of deriving an estimate of the increase in the extension in the case of a passage from z_1 to z_2 reduces to an estimate of the variation of the extension for a passage from the region $D(\Gamma)$ to the region $D(\tilde{\Gamma})$. Since the displacement $z_2 \to z_1$ gives a deflection $|z_2 - z_1| = |\Delta z|$, and the rotation a deflection of the order $|\Delta z|^\alpha$, the deflection of $\tilde{\Gamma}$ from Γ will, in agreement with the conditions of the theorem, be determined by the following conditions:
1) $\tilde{\Gamma}$ lies at a distance from Γ which is not greater than $|\Delta z|^\alpha$;
2) in the neighbourhood of the point z_1, the points of $\tilde{\Gamma}$ lie from those of Γ at distances of the order s^α, where $s = |z - z_1|$.
Introducing an auxiliary plane ω for the estimate of $\delta|f'|$, we obtain in this plane a majorant region of the form $\rho = 1 + |\varphi|^\alpha$ for $\rho \leq 1 + |\Delta z|^\alpha$ and $\rho = 1 + |\Delta z|^\alpha$ at all other points. Thus an estimate for the unknown variation $\Delta|f'|$ can be obtained.

Method of conjugate harmonic functions. By the conditions of the second part of the theorem, the angle α, determining the direction of the tangent to γ, satisfies a Hölder condition with respect to the arc s:

$$|\alpha(s + \Delta s) - \alpha(s)| < k|\Delta s|^\nu;$$

however, since by the first part of the theorem in the case of mapping of $D(\Gamma)$ on to the unit circle the ratio of the corresponding arcs of the curve γ to the arc $\Delta\varphi$ of the circle is bounded along γ from above and below:

$$0 < k_1 < \frac{\Delta s}{\Delta \varphi} < k_2 < \infty,$$

we conclude that $Q = \arg f'$ satisfies along γ a Hölder condition. The quantity Q will satisfy the same condition on the corresponding arc of the circle, if we substitute for z its expression in terms of φ, where $w = e^{i\varphi}$.

Thus, on some arc δ of the unit circle, the harmonic function Q, regular in the circle $|w| < 1$, assumes the values $\alpha = \alpha(\varphi)$ satisfying a Hölder condition. However, in that case, the function

$$P = \log |f'(z, \Gamma)|,$$

conjugate to Q, can be represented in terms of $\alpha(\varphi)$ with the aid of the Poisson integral

$$P(r, \varphi) = \frac{1}{2\pi} \int_0^{2\pi} \frac{2r\alpha(t) \sin (\varphi - t)}{1 - 2r \cos (\varphi - t) + r^2} \, dt,$$

and for points of the circle $|w| = 1$ we obtain

$$\log |f'(z, \Gamma)| = \frac{1}{2\pi} \int_0^{2\pi} \alpha(t) \operatorname{cotan} \frac{\varphi - t}{2} \, dt. \tag{2.3}$$

Thus, $|f'(z)|$ can be expressed with the aid of a singular integral in terms of $\alpha(t)$; however, by a well-known theorem, if $\alpha(t)$ satisfies a Hölder condition, then the integral (2.3) also satisfies this condition, and hence $|f'(z)|$, as a function of φ, also satisfies a Hölder condition. However, since $\Delta s/\Delta\varphi$ is bounded, $|f'(z)|$ satisfies these conditions also as a function of s.

2.2. Narrow strips

The extension at a boundary point of the region $D(\Gamma_0, \Gamma)$ in the case of the conformal mapping

$$w = f(z, \Gamma_0, \Gamma) \tag{2.4}$$

of this region on to the rectilinear strip $0 < v < 1$ depends on the location of the point under consideration and on the entire boundary of the region. As a consequence, the fundamental boundary values problems which involve derivatives at the boundary lead to functional equations. We will now derive several approximate formulae for the determination of the edge extension as a function of the local properties of the boundary curve near a point, where extension occurs. As will be seen below, exact solutions may be constructed on the basis of these formulae.

Consider the mapping (2.4) and, employing hydrodynamical terminology, investigate two infinitely close stream lines: γ_v and $\gamma_{v+\Delta v}$. Select on γ_v a point z_1 and draw through this point the normal to γ_v up to its intersection with $\gamma_{v+\Delta v}$ at the point z_2; let Δ_n be the length of the resulting segment. The ratio $\Delta v/\Delta n$ is the ratio of the amount Δv of flow of fluid in the strip $D(\gamma_v, \gamma_{v+\Delta v})$ to its width. This ratio renders an approximate expression for the flow velocity v in the strip

$$|f'(z, \Gamma_0, \Gamma)| = v \approx \frac{\Delta v}{\Delta n}. \tag{2.5}$$

The formula (2.5) is the simplest hydraulic formula for the determination of the flow velocity through a channel. It is not difficult to make this formula more exact by taking into consideration the following approximate velocity distribution across the section z_1z_2. In fact, the angle α of the flow line is the imaginary part of $\log f'$, whence by virtue of the Cauchy-Riemann equations

$$\frac{\partial \log |f'|}{\partial n} = \frac{\partial \alpha}{\partial s}. \tag{2.6}$$

Introducing a local coordinate system s and n with origin at the centre of the segment z_1z_2, we obtain for the velocity along z_1z_2 from (2.6)

$$\log \frac{V}{V_{av}} \approx k_n \frac{\Delta n}{2},$$

where k_n is the curvature of the stream line with corresponding signs. Setting $e^{k_n \Delta n/2} \approx 1 + (k_n \Delta n)/2$, we can rewrite the last formula in the form

$$V \approx V_{av}\left(1 + k_n \frac{\Delta n}{2}\right). \tag{2.7}$$

The formula (2.7) has been derived under the assumption that the stream lines γ_v and $\gamma_{v+\Delta v}$ are analytic; using variational principles, one can greatly extend the region of application of this formula and give an estimate of the remainder term. Assume that γ_v and $\gamma_{v+\Delta v}$ satisfy the following conditions:

1°. the curvatures k of both curves exist and satisfy the Hölder condition

$$|k(s + \Delta s) - k(s)| < \lambda_1 |\Delta s|^{\nu}; \tag{2.8}$$

2°. the difference of the curvatures and the inclinations of γ at the points z_1 and z_2 are small together with the width of the strip:

$$|k_2 - k_1| < \lambda_3 \Delta n, \ |\alpha_2 - \alpha_1| < \lambda_2 \Delta n; \tag{2.9}$$

3°. the width of the strip and the amount of flow are subject to the bound

$$\left| \log \frac{\Delta v}{\Delta n} \right| < \lambda_0. \tag{2.10}$$

Under these assumptions, draw through the point z_1 a circle C_1, which is in contact with γ_v, and through z_2 a circle C_2, concentric with C_1, and replace the curves γ_v and $\gamma_{v+\Delta v}$ by the circles C_1 and C_2. The mapping of the annular multi-sheeted region $D(C_1, C_2)$ on to the horizontal strip has the form

$$w = iA \log (z - z_0),$$

where z_0 is the common center of C_1 and C_2 and A is the real constant:

$$A = \frac{\Delta v}{\log \left| \dfrac{z_2 - z_0}{z_1 - z_0} \right|} = \frac{\Delta v}{\log \left(1 + \dfrac{\Delta n}{r_1} \right)}, \ r_1 = |z_1 - z_0|.$$

The problem of deriving an estimate of the accuracy of (2.7) reduces to the estimate above of the variation of the extension in the case of a transition from $D(\gamma_v, \gamma_{v+1})$ to the ring $D(C_1, C_2)$. Under the conditions 1°.–3°. assumed above, we obtain for the error R a quantity of order Δn^2.

Continuing in the same area of ideas, it is not difficult to construct a formula and to present an estimate of its accuracy with a supplementary evaluation of the curvature of $\gamma_{v+\Delta v}$ and of the inclination of this curve at the point z_2 with respect to γ_v. Let k_2 be the curvature of $\gamma_{v+\Delta v}$ at z_2 and θ the angle between the tangent to $\gamma_{v+\Delta v}$ at z_2 and the tangent to γ_v at z_1; then one finds for $|f'|$ at z_1 the expression

$$|f'(z, \gamma_v, \gamma_{v+\Delta v})| = \\ = \frac{\Delta v}{\Delta n}\left\{1 + \frac{\Delta n}{3} k_2 + \frac{\Delta n}{6} k_1 + \frac{\Delta n^2}{12} k_2^2 + \tfrac{1}{3}\theta^2 + R\right\}. \quad (2.11)$$

The principal term of this formula is derived by replacing $D(\gamma_v, \gamma_{v+\Delta v})$ by a circular crescent shape bounded by the circles C_1 and C_2', touching $\gamma_{v+\Delta v}$ at the point z_2. The remainder term R can be estimated by the earlier method, provided one introduces additional limitations on the curvature and the angle θ.

Thus, we assume, in addition, that

$$|k_2 - k_1| \leq k' \Delta n \quad (2.12)$$

and that the second derivatives of the curvatures k_1 and k_2 are bounded:

$$\left|\frac{d^2 k_1}{ds^2}\right| \leq k'', \quad \left|\frac{d^2 k_2}{ds^2}\right| \leq k''. \quad (2.13)$$

Under these conditions, we obtain for R an estimate of the form

$$|R| \leq A k'' \Delta n^3, \quad (2.14)$$

where A is some constant.

Note the very important particular case of (2.11) when Γ_0 coincides with the x-axis; we obtain for an arbitrary point of the curve Γ

$$|f'| = \frac{\Delta v}{\Delta n}(1 + \tfrac{1}{3}\Delta n k + \tfrac{1}{3}\theta^2 + R_1). \quad (2.15)$$

As before, assuming the angle θ and the curvature k to be small, one can rewrite (2.15) exactly, apart from small higher order terms, in the form

$$|f'| = \frac{\Delta v}{y}\left(1 + \tfrac{1}{3}yy'' + \tfrac{1}{3}y'^2 + R_2\right). \tag{2.16}$$

Estimates for the remainder terms R_1 and R_2 can be derived as functions of the boundary values of y''' and $y^{\prime\mathrm{v}}$. In fact, if $|y'''(x)| \leq k'''$, then

$$|R| \leq k'''h^2, \tag{2.17}$$

and if $|y^{\prime\mathrm{v}}(x)| \leq k^{\prime\mathrm{v}}$, then

$$|R| \leq k^{\prime\mathrm{v}}h^3. \tag{2.18}$$

If we assume, in addition, that

$$|y'| < k'h^{\frac{3}{2}}, \quad |y''| < k''h, \quad |y'''| < k'''h^{\frac{1}{2}}, \tag{2.19}$$

then (2.16) can be simplified to read

$$|f'| = \frac{\Delta v}{y}\left(1 + \tfrac{1}{3}yy'' + R\right), \tag{2.20}$$

where an estimate of R is given by

$$|R| < Ah^{\frac{5}{2}} \tag{2.21}$$

with A depending only on the constants k.

Next, we present a supplementary statement referring to the evaluation of the remainder term. If we map the crescent shaped region $D(C_1, C_2')$ on to the strip $0 < \eta < h$, then Γ_0 and Γ become the curves

$$\begin{aligned} \eta &= \eta_0(\xi) = A_0\xi^3 + B_0\xi^4 + \cdots, \\ \eta &= \eta_1(\xi) = h + A_1(\xi - \xi_0)^3 + B_1(\xi - \xi_0)^4 + \cdots, \end{aligned} \tag{2.22}$$

where $\xi_0 = k'h^2$. We will now perform similar expansions of both planes ω and w in the ratio $1/h$; in that case the curves (2.22) become the lines

$$\begin{aligned} y &= y_0(x) = A_0h^2x^3 + B_0h^3x^4 + \cdots, \\ y &= y_1(x) = 1 + A_1h^2(x - x_0)^3 + B_1h^3(x - x_0)^4 + \cdots, \end{aligned} \tag{2.23}$$

where $x_0 = k'h$. Then the value of f' does not change, and our problem reduces to the estimate of the deviation from unity of the derivative in the case of the mapping of the strip $y_0(x) < y < y_1(x)$ on to the strip $0 < v < 1$.

Letting

$$z = F(w) = x(u, v) + iy(u, v)$$

denote the function which maps in an invertible manner the strip $0 < v < 1$ on to the region $y_0(x) < y < y_1(x)$, we will have at the point $z = 0$ or $w = 0$ of interest here

$$|F'(0)| = \frac{1}{|f'(0)|} = \frac{\partial y(u, v)}{\partial v}.$$

The function $y(u, v)$ is a harmonic function which assumes on the boundaries of the strip $0 < v < 1$ the values

$$y(u, 0) = y_0[x_0(u)], \quad y(u, 1) = y_1[x_1(u)],$$

where $x_0(u)$ and $x_1(u)$ are functions expressing the correspondence on the boundaries in the case of the transformation of the strip. By virtue of the smoothness of the boundary [1] and the repre-

---

[1] One has the following

THEOREM. *If Γ_0 is such that the angle of inclination α of its tangent as a function of the arc length s has an n-th order derivative, satisfying a Hölder condition, then $f'(z, \Gamma_0, \Gamma)$ exists along Γ_0 and has an n-th order derivative with respect to s which satisfies a Hölder condition with the same index.*

Proof. Construct through the point z_0 of the curve Γ_0 a curve γ given by a polynomial which has with Γ_0 contact of order n and map the region $D(\gamma_0, \Gamma)$ on to the strip $0 < \eta < 1$. The problem of the existence of $f^{(n)}(z, \Gamma_0, \Gamma)$ reduces to that of the existence of $\partial p/\partial v$, where p is a harmonic function assuming on $v = 0$ the values $y = y(x)$ with $y(0) = y'(0) = y''(0) = \cdots = y^{(n-1)}(0) = 0$. The existence of $\partial p/\partial v$, however, follows from the convergence of the integral $\int_{-\infty}^{+\infty} y(t)\,dt/(\cosh^2 t/2)$. The Hölder condition for $\partial p/\partial v$ can be derived from the n-th order proximity of the region obtained by displacement of D by Δs.

sentation (2.23), we have

$$y(u, 0) = y_0[u + \alpha_0 u^2 + \cdots] = A_0' h^2 u^3 + B_0' h^3 u^4 + \cdots,$$
$$y(u, 1) = y_1[u + \alpha_1 u^3 + \cdots] = A_1' h^2 (u - u_0)^3 + B_1' h^3 (u - u_0)^4 + \cdots$$

In order to obtain the estimate required, one has still to use the representation of $\partial p/\partial v$ in terms of the boundary values of y given by the well-known formula (cf., for example, M. A. Lavrent'ev and B. V. Shabat, 1958, p. 362):

$$\frac{\partial y}{\partial v} = 1 + \frac{1}{2\pi} \int_{-\infty}^{+\infty} \frac{y(t, 0)\, dt}{\sinh^2 \frac{t}{2}} + \frac{1}{2\pi} \int_{-\infty}^{+\infty} \frac{y(t, 1)\, dt}{\cosh^2 \frac{t}{2}}.$$

The Dirichlet Problem. We have considered the approximate solution of problems of conformal mapping of narrow strips on to rectilinear strips. This problem, in principle, is equivalent to the Dirichlet Problem for such strips: By mapping a strip on to a rectilinear strip, we reduce the Dirichlet Problem for a curvilinear strip to that for a rectilinear strip. However, taking into consideration the narrowness of the strip and the smoothness of the boundary conditions of the Dirichlet problem, one can, as in the case of conformal mapping, replace the functional approximately by a function.

Thus, let there be given on the boundaries of the rectilinear strip $0 < y < h$ the values of the harmonic function $u(x, y): u(x, 0) = y_0(x)$, $u(x, h) = y_1(x)$; for this purpose let y_0 and y_1 satisfy the conditions

$$|y_1(x) - y_0(x)| < k_0 h, \quad |y_{0,1}'| < k_1, \quad |y_{0,1}''| < k_2, \quad |y_{0,1}'''| < k_3. \tag{2.24}$$

It is required to find $\partial u/\partial y$ at an arbitrary point of the boundary.

Select the first approximation to the solution of this problem in the form

$$\frac{\partial u}{\partial y} \approx \frac{y_1(x)}{y_0(x)}. \tag{2.25}$$

This formula can be made more exact by replacing $y_0(x)$ and $y_1(x)$ by contacting parabolae; in that case one obtains along the x-axis

$$\frac{\partial u}{\partial y} = \frac{y_1(x)}{y_0(x)} [1 + a y_0''(x) + b y''(x)] + R, \tag{2.26}$$

where a and b are some absolute constants and the remainder term R is of third order of smallness in h.

2.3. Behaviour of the extension at points of maximum inclination and extreme curvature

Next, we establish several propositions referring to the behaviour of the extension $|f'(z, \Gamma_0, \Gamma)|$ at points of Γ in dependence on the behaviour of Γ at these points. In order to simplify the analysis, let $p(s)$ denote the function $\log |f'(z, \Gamma_0, \Gamma)|$ on Γ as a function of the arc length s:

$$p(z) = p(s) = \log |f'(z, \Gamma_0, \Gamma)|, \ z \subset \Gamma. \tag{2.27}$$

THEOREM 2.2. *Let Γ_0 and Γ be smooth curves given by the single-valued functions* $y = y_0(x)$, $y = y_1(x)$. *If the gradient $y_1'(x)$ attains at the point z_0 of Γ an absolute maximum, where this maximum exceeds* $\max y_0'(x)$, *and if at this point $p'(s)$ exists, then*

$$p'(s) < 0. \tag{2.28}$$

Proof. In fact, the angle of inclination $\alpha = \arg f'$ and the function p are conjugate harmonic functions; consequently, along Γ,

$$\frac{\partial \alpha}{\partial n} = - \frac{\partial p}{\partial s}.$$

Since by assumption the function α attains at z_0 an absolute maximum, it follows that $\partial\alpha/\partial n > 0$, which is equivalent to (2.28).

In the case of applications, the following generalization of the theorem proved will be essential:

THEOREM 2.3. *Under the conditions of Theorem 2.2 assume that the absolute maximum of the function α is attained on the* *segment* $\gamma \subset \Gamma$ *with ends* s_1 *and* s_2; *in addition, assume that* Γ *has bounded curvature* [1]. *Then we have at the point* s_0

$$\lim_{\Delta s \to 0} \frac{p(s_0 + \Delta s) - p(s_0)}{\Delta s} < 0. \tag{2.29}$$

Proof. 1) For the proof assume that (2.29) is untrue, i.e., that

$$\lim_{\Delta s \to 0} \frac{p(s_0 + \Delta s) - p(s_0)}{\Delta s} \geq 0$$

and consider the mapping

$$\omega = \log |f'| + i\alpha.$$

In the case of this transformation, the neighbourhood of the points s_0 where the maximum occurs becomes part of a right angle with vertex at the point $p(s_0)$, $\alpha_0 = \max \alpha$, and sides parallel to the p-axis (to the right) and α-axis (downwards, cf. fig. 6). However, in the case of such a transformation, the variation of α along the arc Δs will be of order $\sqrt{\Delta s}$ which contradicts the assumption that the curve Γ is to have bounded curvature.

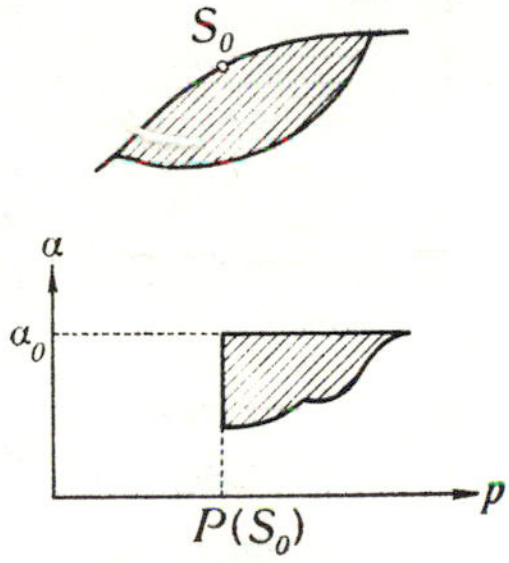

Fig. 6

2) We can also proceed in a different manner. By virtue of the assumption of bounded curvature, the function p will satisfy the condition

$$|p(s + \Delta s) - p(s)| < k|\Delta s| \log \frac{1}{|\Delta s|}.$$

On the other hand, in the neighbourhood of the point s_0, we have a harmonic function α which is constant along $s_1 s_2$ and satisfies everywhere a Hölder condition. Such a function cannot render a mapping on to a region with an angular point, because

[1] It is probably sufficient to demand that the function α is to satisfy a Hölder condition.

then the increment of α or p will be of order $\sqrt{\Delta s}$ which is impossible.

NOTE. Theorems which are analogous to the ones just proved hold also in the case when α attains an absolute minimum; then all the inequalities are inverted.

We proceed now to the study of the behaviour of $p(x)$ at points of Γ where Γ attains maximum curvature.

THEOREM 2.4. *Let the curves Γ_0 and Γ satisfy the following conditions:*

1°. *they are defined by single-valued functions $y = y_0(x)$, $y = y_1(x)$, where*

$$0 < k_0 < y_0(x) < y_1(x) < k_0'; \tag{2.30}$$

2°. *the gradients α of the curves are bounded:*

$$|y_0'(x)| < k_0', \; |y_1'(x)| < k_1'; \tag{2.31}$$

3°. *the curvatures of the curves are bounded:*

$$|k_0(x)| = \left|\frac{d\alpha_0}{ds}\right| \leq k_0, \quad |k_1(x)| = \left|\frac{d\alpha_1}{ds}\right| \leq k_1; \tag{2.32}$$

4°. *the curve Γ contains an arc γ of a circle with curvature $1/r$.*

Under these conditions, assuming k_1 and k_1' to be fixed, for $1/r$ sufficiently large (depending only on k_0 and k_1), we will have at every point of the arc γ

$$\frac{d^2p}{ds^2} > \frac{k}{r^2} \quad \text{or} \quad \frac{d^2p}{ds^2} < -\frac{k}{r^2}, \tag{2.33}$$

where the signs $>$ or $<$ depend on whether γ is convex or concave with respect to $D(\Gamma_0, \Gamma)$; for sufficiently small r the constant k depends only on k_0, k_1, k_0', k_1' and $p = \log|f'(z, \Gamma_0, \Gamma)|$. Under those conditions, we have at the ends of the arc γ

$$\overline{\lim_{\Delta s \to 0}} \frac{p(s + \Delta s) - p(s)}{\Delta s^2} \begin{cases} > 0, \text{ if } y'' < 0 \text{ on } \gamma, \\ < 0, \text{ if } y'' > 0 \text{ on } \gamma. \end{cases} \tag{2.34}$$

Proof. In order to derive the inequalities (2.33) and (2.34), we will construct upper bounds for their left-hand sides. For this purpose, we require certain supplements to the formulae for the variations of p, established earlier. It follows from (2.11) that, if we bring about a deformation of the curve Γ in the neighbourhood of the point z_1, replacing $y = y(x)$ by $y = \tilde{y}(x)$, where $\tilde{y}(x) \geq y(x)$ and $\tilde{y}(x)$ differs from $y(x)$ only in a small neighbourhood of x_1, then we will have exactly, apart from small, higher order quantities,

$$|f'(z, \Gamma_0, \tilde{\Gamma})| = |f'(z, \Gamma_0, \Gamma)| \left\{1 + \frac{\pi}{4} \frac{|f'(z_1, \Gamma_0, \Gamma)|}{\sinh^2(u - u_1)} \sigma\right\}$$

or

$$\delta p(s) = \frac{A}{\sinh^2(u - u_1)} \sigma,$$

where σ is the "area" of variation

$$\sigma = \int_{-\infty}^{+\infty} [\tilde{y}(x) - y(x)]\, dx,$$

u_1 and u are the abscissae of the points of the straight line $v = 1$ which correspond in the case of the mapping $w = f(z, \Gamma_0, \Gamma)$ to the points z_1 and z of the curve Γ.

If we select, side by side with the point z_1, a point z_2 and induce in the neighbourhoods of these points variations with areas σ_1 and σ_2, we obtain for the corresponding variations of p

$$\delta p(s) = \frac{A_1}{\sinh^2(u - u_1)} \sigma_1 + \frac{A_2}{\sinh^2(u - u_2)} \sigma_2.$$

If σ_1 and σ_2 have the same signs, the function δp has between the points x_1 and x_2 a single extreme value, a maximum for $\sigma > 0$, a minimum for $\sigma < 0$. For any fixed x_1, σ_1, x_2 and x_0, $x_1 < x_0 < x_2$, one can select σ_2 such that this extreme value occurs at the point x_0; at this point, we will have

$$\frac{d}{du} \delta p(s) = \frac{1}{V} \frac{d}{ds} \delta p(s) = 0 \tag{2.35}$$

and, in addition,

$$\frac{d^2}{du^2}\,\delta p \left.\begin{matrix} \geq 0, \text{ if } \sigma > 0, \\ \leq 0, \text{ if } \sigma < 0. \end{matrix}\right. \tag{2.36}$$

It follows likewise from (2.35) and (2.36) that the inequality (2.36) applies at the point x_0, if we differentiate with respect to s instead of with respect to u:

$$\frac{d^2}{ds^2}\,\delta p \left.\begin{matrix} \geq 0, \text{ if } \sigma > 0, \\ \leq 0, \text{ if } \sigma < 0. \end{matrix}\right. \tag{2.37}$$

It follows directly from the inequality (2.37) that in the case of an expansion of the region $D(\Gamma_0, \Gamma)$ by a corresponding deformation of the curve Γ outside γ, we can increase our functional d^2p/ds^2, i.e., we have

$$\delta \frac{d^2p}{ds^2} > 0, \tag{2.38}$$

and in the case of a construction of the region $D(\Gamma_0, \Gamma)$ the same functional can be decreased. It follows directly from (2.38) that, in order to derive an upper bound for d^2p/ds^2, one may select a region of a special form, bounded by arcs of circles and straight lines (fig. 7); the required inequality can then be derived directly by computations. However, in this manner, there arise so extensive, largely computational difficulties, that we will fall back on further simplifications.

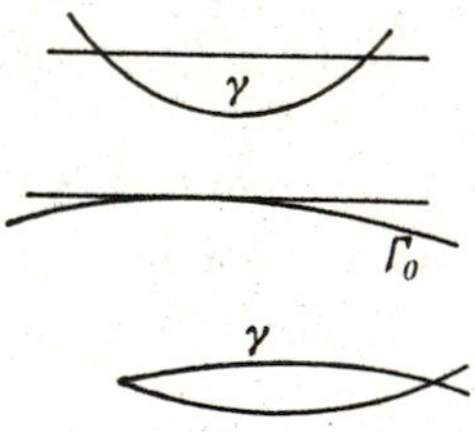

Fig. 7

In the case when the curvature $1/r$ is large and the width of the strip $k_1' - k_0'$ is fixed, the influence of the curve Γ_0 on the derivative under consideration will be small. An increase in curvature is equivalent to a decrease of the width of the strip; instead of a strip, one can consider the regions $y < y(x)$ and $v < 0$. Then one can construct regions which can be mapped in an elementary manner on to the half-plane. In order to derive a lower bound for d^2p/ds^2, one can select as such a majorant region the region bounded by the semi-circle $|z| = r$, $y > 0$ and the x-axis. Then one can take as mapping function $w = \frac{1}{2}(z + 1/z)$ and with $z = e^{i\varphi}$ we obtain $u = \cos\varphi$, whence $p = \log\sin\varphi$ and

$$\frac{d^2p}{ds^2} = \frac{d^2p}{d\varphi^2} = \frac{1}{\sin^2\varphi} > 1.$$

The second part of the theorem can be proved in an analogous manner and therefore we will devote no time to it here.

CHAPTER III

HYDRODYNAMIC APPLICATIONS

Proceeding to applications of these principles, we begin with classical problems of the theory of stream line flow of an ideal fluid. In order to avoid burdening of the study with auxiliary manipulations, we consider the problem of constructing the flow with a free stream line in a formulation which is more suitable for the present method, and only note the other problems and the extensions of the method which are necessary for their solution.

3.1. Stream line flow

Let there be given a curve Γ_0: $y = y_0(x) \geq 0$, $|y_0'(x)| < k'$, $|y_0''(x)| < k''$, located above the x-axis for $0 < x < 1$ and coinciding with the x-axis outside this segment. Let it be required to construct the steady flow of an ideal fluid which is symmetric about the x-axis such that

1°. the amount of fluid is equal to 2;
2°. the pressure along the free surface is constant;
3°. the velocity of the flow at infinity is equal to 1;
4°. one has stream line flow around the contour $y = \pm\, y_0(x)$.

The hydrodynamic problem under consideration is equivalent to the following problem:

Construct a curve Γ: $y = y(x) < y_0(x)$ such that in the case of the transformation $w = f(z, \Gamma_0, \Gamma)$ of the flow region $D(\Gamma_0, \Gamma)$ on to a strip of unit width $0 < v < 1$ one would have along Γ

$$V = |f'(z, \Gamma_0, \Gamma)| = 1. \tag{3.1}$$

The last condition follows from the assumptions relating to the velocity of the flow at infinity and the constant pressure on Γ

(by virtue of Bernoulli's equation, the pressure $p = C - \rho V^2/2$, where C and ρ are constants).

In order to find the solution of the problem above we will seek a minimum of the functional

$$I(\Gamma) = \max |\log |f'(z, \Gamma_0, \Gamma)|| \tag{3.2}$$

in the compact class of admissible curves which are characterized by the conditions:

1°. $k_0 \leq y(x) - y_0(x) \leq k_0'$;

2°. $|y'(x)| \leq \max |y_0'(x)| + 1 = k_1$;

3°. $|k| = \left| \dfrac{y''}{(1 + y'^2)^{\frac{3}{2}}} \right| \leq \max \left| \dfrac{y_0''}{(1 + y_0'^2)^{\frac{3}{2}}} \right| + M = \dfrac{1}{r}$.

We will dispose of the constants in the following manner. In order to fix the constants k_0 and k_0', we draw two curves Γ_1: $y = y_0(x) + H$, and Γ_2: $y = y_0(x) + h$. Obviously, for Γ_1, we have $p(\Gamma_1) > 0$ everywhere for H sufficiently large and, for Γ_2, we find $p(\Gamma_2) \to -\infty$ as $h \to 0$; consequently, we have for sufficiently small h that $p(\Gamma_2) < 0$.

The values of h and H, for which the inequalities above are satisfied, will be selected for k_0 and k_0'. We select the constant M somewhat larger, in order to fulfill the conditions of Theorem 2.4.

By virtue of the basic principle and this theorem, the functional $I(\Gamma)$ and the class of admissible curves satisfy the conditions of Theorem 2.4. Hence we conclude that there exists in the class of admissible curves selected a solution for which

$$I(\Gamma) = 0.$$

By virtue of the basic variational principle, this solution is unique and stable: if $\tilde{\Gamma}$ deviates from Γ by more than ε, then $I(\tilde{\Gamma}) > \varepsilon$.

In fact, we can obtain the minimum variation of V, if we deform the curve Γ everywhere in the direction of its normal by ε; however, since $V = 1$ everywhere on Γ, we conclude then that

$$\delta I(\Gamma) = \varepsilon.$$

3.2. Generalizations

1) With hardly any change of the reasoning above we may obtain a somewhat more general result by replacing the condition (3.1) in the problem under consideration by the more general condition

$$V = 1 + \lambda\varphi(x, y), \tag{3.3}$$

where φ is a positive function with partial derivatives up to and including the second order, which are continuous and bounded, and λ is a parameter. In hydrodynamic language, we obtain for $\varphi(x, y) \equiv y$ the motion in a gravity field, and such a formulation contains the entire theory of waves. We will consider below one problem of this type. At the present time we confine ourselves to one remark: the solution exists and is unique and stable for all sufficiently small values of the parameter λ.

An estimate of the first value of the parameter λ for which the problem can be solved may be obtained. It is sufficient for the applicability of our method that in the case of a variation of Γ the principal term in the variation of the function $\{V-\lambda\varphi(x, y)\}$ should be the variation in V.

2) We present now the example of a problem which may be solved directly by the method under consideration. Let there be given two sufficiently smooth curves Γ_0 and Γ_1 which can be represented by single-valued functions

$$y = y_0(x), \ y = y_1(x), \ |x| < \infty,$$

where

$$\begin{gathered} k_0 < y_1(x) - y_0(x) < k_0', \\ |y_{0,1}'(x)| < k_1, \ |y_{0,1}''(x)| < k_2; \end{gathered} \tag{3.4}$$

it is required to construct a curve Γ: $y = y(x)$, $y_0(x) < y(x) < y_1(x)$ such that along this curve

$$|f'(x, \Gamma_0, \Gamma)| = F\{|f'(x, \Gamma, \Gamma_1)|, x, y\}, \tag{3.5}$$

where $F(V, x, y)$ is a given function.

We can apply the method directly to the case when the function $F(V, x, y)$ increases as a function of $V(0 < k < \partial F/\partial V)$, depends sufficiently weakly on y (i.e., $\partial F/\partial y$ is sufficiently small) and, finally, has uniformly bounded partial derivatives with respect to all arguments.

For particular forms of the function F the last problem lends itself to simple hydrodynamic interpretations: the curve Γ is the interface between two flows with different densities.

3.3. Stream line flow with detachment

Consider now the following classical problem (Kirchhoff). Let there be given an arc γ: $y = \pm \varphi(x)$, $0 \leq x \leq a$, symmetric about the x-axis; let the function $\varphi(x)$ be single-valued and have bounded curvature. Let γ be placed in the flow of an ideal fluid with given velocity at infinity. It is required to construct the flow, assuming that free stream lines originate from the ends of the arc and that the flow is symmetric about the x-axis (fig. 8).

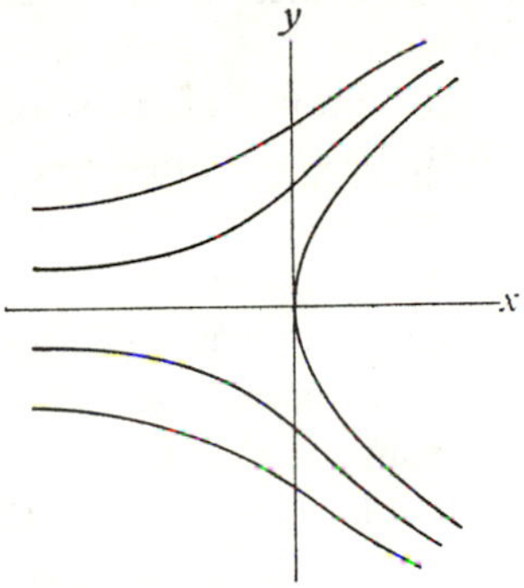

Fig. 8

In terms of the theory of functions of a complex variable, this problem may be formulated in the following manner:

It is required to construct a curve Γ: $y = y(x)$, such that

1) $y(x) \equiv 0$ for $x \leq 0$;
2) $y(x) \equiv \varphi(x)$ for $0 \leq x \leq a$;
3) the identity $|f'(z, \Gamma)| \equiv 1$ holds true along the curve Γ for $x > a$.

The theorems of the preceding chapter do not permit direct application of the method to this problem. The reason is that, as is shown by the example of the stream line flow around a segment of a straight line, at the point where the free stream lines begin the curvature of Γ may become infinite, i.e., the solution of this problem may not be contained in the class of admissible curves with bounded curvature.

Nevertheless, one may obtain the solution of this problem by a variational method, modifying the preceding reasoning by the following additions:

1) we determine first of all some modified problem, assuming that γ has rectilinear circuits around the ends and that the free stream line passes at a finite distance from the rectilinear segment in the flow;

2) we select a somewhat more general class of admissible curves γ than before: the condition of boundedness of the curvature will be replaced by the condition that $y'(x)$ satisfies a Hölder condition and that for $x \to 0$ the curvature of γ may not tend to infinity faster than ν/x, where ν is a sufficiently small constant.

It seems to the author that this method may be made still more flexible by inessential improvements and that it will then render solutions of considerably more general problems of flows with free stream lines.

One may seek the solution of the problem in the case when on the free stream-line a sufficiently general relationship between the velocity and the position of the point is given; one may consider likewise the case of non-symmetry about the x-axis as well as the case of free stream line flow in two media with different densities.

From the point of view of application, special interest attaches to problems relating to the study of the actual area of cavitation. The fact is that the above classical scheme of Kirchhoff with free stream lines going to infinity is not realized in practice, i.e., the stream closes at a finite distance from the body in the flow. As a consequence, a number of other schemes have been proposed.

Riabouchinsky's scheme is to arrange with the body Γ_0 in the flow a mirror image Γ_1 (fig. 9) and to seek curves γ_+ and γ_-, joining the upper and lower ends of Γ_0 and Γ_1, respectively, such that for the stream line flow around the region, bounded by Γ_0, γ_-, Γ_1 and γ_+, the velocity maintains along γ_- and γ_+ a constant value.

Much later, a new scheme was proposed (Efros) in which it

was assumed that a stream line δ of the fluid runs along the axis of symmetry and further on to Γ_0, reverts by 180°, turns back to Γ_0 along the axis of symmetry and then goes to infinity (fig. 10). The amount of flow along the free stream line δ and its

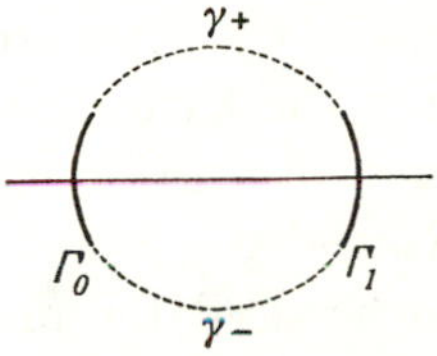

Fig. 9

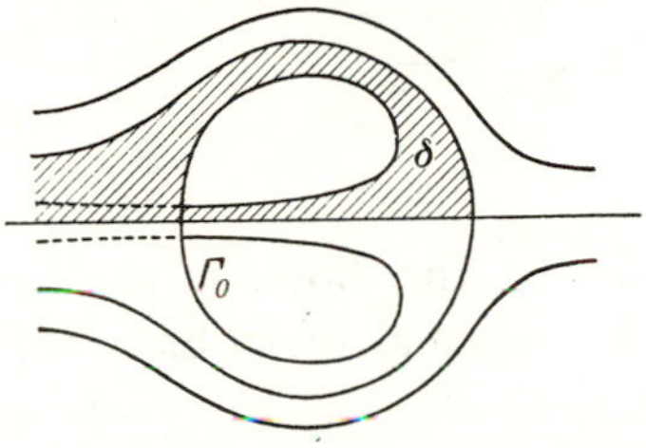

Fig. 10

internal boundary is determined by the condition that on this boundary the velocity of the flow must assume a given value. Efros' scheme does not introduce an auxiliary body, but the turning back towards the body of the stream line must occur mathematically on a second sheet of the Riemannian surface which lacks physical significance.

Both schemes under consideration, in spite of their artificiality, render (for the analysis of the velocity distribution along Γ_0) better correlation with experiment than Kirchhoff's scheme. We will still note another scheme which is not very different from Efros', but has been deprived of the physical inconsistencies (it was proposed by the author and presented in a seminar in Novosibirsk in 1958; for the case when Γ_0 is a segment of a straight line, the problem has been solved completely by Pykhteev). We assume that behind the body in the flow, i.e. the arc Γ_0, there are rings δ filled with liquid. The ring δ^+ is bounded from outside by a curve which consists of the stream line γ^+ passing through the upper end of Γ_0, the upper half of the arc Γ_0 and a segment of the x-axis; on the inside, δ^+ is bounded by a curve γ_0 (fig. 11). The conditions determining the basic flow of the fluid and the flow of the liquid in the ring comprise the following:

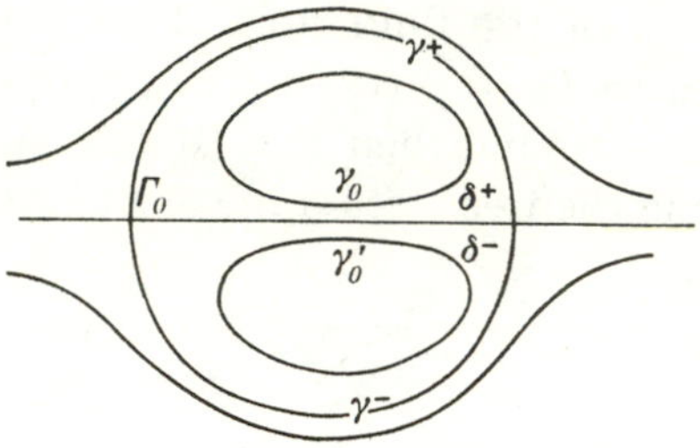

Fig. 11

1) along γ_0, the velocity of the fluid in the ring must have a given constant value;

2) along $\gamma^\pm$, the velocity of the fluid in the ring must coincide with the velocity of the basic flow around the arc Γ_0 augmented by the arcs $\gamma^\pm$. Assuming the basic flow to be irrotational, the flow in the rings δ^+ and δ^- can be assumed to be irrotational or to have constant vorticity.

The methods studied above may also be used for problems arising in the theory of explosions. The fact is that in a region close to the explosion, the elastic and viscous forces of practically any medium (grounds, metals, etc.) are small compared with the inertia forces; on the other hand, with the distance from the centre of the explosion, the inertia forces rapidly decrease and, thus, at a certain distance these forces are small compared with the resistance forces. Hence it is natural to deal with models for the computation of a first approximation in which the entire medium is divided into two parts: one (close to the centre of the explosion) is considered to be an ideal fluid, the remainder a perfectly rigid body. I will present now an accurate description of one of those models studied first for a very simple case by Kuznetsov.

Let the medium fill that part of space located below the curve $\Gamma : y = y(x)$ (we restrict ourselves here, as above, to the plane case) and let there have acted on the medium along the section AB of the curve Γ an impulse $J = J(x)$; it is required to determine the velocity distribution in the medium as a function of time (fig. 12).

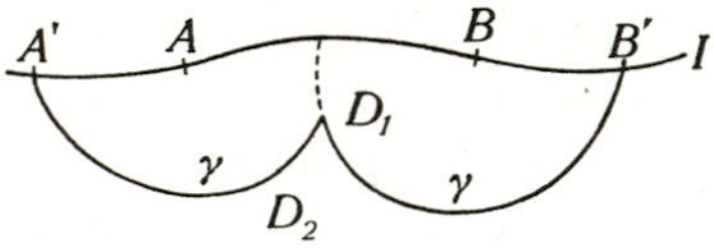

Fig. 12

In correspondence with the statements above, we divide the medium by a line γ (with ends A' and B' on Γ) into a region D_1, which has the arc AB as its boundary, and into a region D_2, which is complementary to D_1 and lies below Γ. We assume the part of the medium contained in D_1 to be an ideal fluid, and that contained in D_2, a perfectly rigid body. The problem of determining the velocity distribution in D_1 for a given impulse on AB reduces immediately to a well-known mixed problem. Denoting the unknown velocity potential by φ, we obtain for φ the following conditions:

$$\varphi = J(x) \text{ on } AB,$$

$\varphi = 0$ on that part of Γ which is contained in the boundary of D_1,

$$\partial\varphi/\partial n = 0 \text{ everywhere on } \gamma \text{ (flow condition).}$$

Our problem is then that, given a certain supplementary condition of a physical character, one can determine γ; it is natural to use as such a condition the requirement that everywhere in D_1 the inertia forces, i.e., the relative velocities should be larger than some constant, determined by the strength of the material. Under these assumptions, our problem leads to the determination of γ such that the flow arising in D_1 under the influence of the impulse J has a constant velocity along γ.

3.4. Wave motions of a fluid

The problem of stationary plane waves in a channel of finite depth can be formulated in the following manner.

It is required to construct in the z-plane a periodic curve

$$\Gamma : y = y(x),\ y(x + \omega) = y(x),\ y(0) = h,\ y''(0) < 0 \quad (3.6)$$

symmetric with respect to the y-axis such that for conformal transformation of the region $D(\Gamma_0, \Gamma)$, where Γ_0 coincides with the x-axis, on to the strip $0 < v < H$ one has along Γ the relation

$$\frac{\rho}{2}\,|f'(z, \Gamma_0, \Gamma)|^2 + \rho g y = C; \quad (3.7)$$

here ρ is the density of the fluid and g the gravitational acceleration. The constants h, H, C are assumed to be given and the number ω is subject to determination.

Using a similarity principle, the number of parameters of the problem can be reduced to two (by linear transformations of the z- and w-planes, we may reduce the problem to the case when $h = H = 1$) and (3.7) assumes the form

$$I(x, \Gamma) = |f'(z, \Gamma_0, \Gamma)|^2 + \lambda y = a, \tag{3.8}$$

where λ and a are the parameters of the problem.

The variational principles relating to mappings of curvilinear strips on to the rectilinear strip permit direct determination of the minimum value of λ beginning with which non-trivial solutions are possible.

For $\lambda \leq 2$, the solution of the problem is possible only for $a = 1 + \lambda$, and this solution corresponds to translatory flow with the potential $w = z$.

In fact, we assume, on the contrary, that beside the solution $w = z$ there exists still a solution with a free surface $y = y(x)$, $y(x) \not\equiv 1$, and potential $w = f(z, \Gamma_0, \Gamma)$. By virtue of Lindelöf's principle (Theorem 1.3) and the condition (3.6), we will have at the point $z = i$

$$|f'(i, \Gamma_0, \Gamma)| < 1;$$

substituting this into (3.8), we find that for our flow $a \leq 1 + \lambda$. On the other hand, let x_0 be the point of a minimum of the function $y = y(x)$ with $y_0 = y(x_0)$; by virtue of the same principle, we will have at the point $z_0 = x_0 + iy_0$

$$|f'(z_0, \Gamma_0, \Gamma)| > \frac{1}{y_0},$$

whence, by (3.8),

$$a = I(x_0, \Gamma) > \frac{1}{y_0^2} + \lambda y_0.$$

The right-hand side of this inequality attains as a function of y_0 a minimum $(3\lambda/2)(2/\lambda)^{\frac{1}{3}}$ at the point $y_0 = (2/\lambda)^{\frac{1}{3}}$, whence $a > (3\lambda/2)(2/\lambda)^{\frac{1}{3}}$. Combining this result with the inequality $a \leq 1 + \lambda$, derived earlier, we arrive at a contradiction for $\lambda \leq 2$, and hence the assertion is proved.

The construction and study of wave motions (for $\lambda > 2$ and for corresponding values of a) have formed the subject of many investigations since the beginning of the last century. The majority of these are concerned with approximate methods and, in the first place, with the linearized problem.

The first studies based on the exact theory have occurred in our own times: Villat reduced the problem to a singular integral equation, Levi-Civitá and Nekrasov gave exact solutions for waves with sufficiently small amplitudes. The methods of Nekrasov and Levi-Civitá were based on the expansion of the kernel of the integral equation in terms of a small parameter; the solution was obtained in the form of series the convergence of which was proved for sufficiently small values of the parameter. Leray, Weinstein, Kravtchenko, Gerber and others have developed a new, topological method for the study of wave motions.

I will present here a method based on variational principles. For this method, a preliminary study of linear and non-linear approximate solutions is essential.

3.5. The linear theory of waves

In the exact formulation of wave theory, we have the non-linear boundary condition (3.8) which may be written down in terms of harmonic functions. Taking as independent variables the components of the complex potential u and v and denoting by $z = F(w) = x(u, v) + iy(u, v)$ the function which is the inverse to $w = f(z, \Gamma_0, \Gamma)$, we will have

$$|f'(z, \Gamma_0, \Gamma)|^2 = \frac{1}{|F'(w)|^2} = \frac{1}{\left(\dfrac{\partial x}{\partial u}\right)^2 + \left(\dfrac{\partial y}{\partial u}\right)^2}.$$

Then (3.8) may be written for $v = 1$ in the form

$$\frac{1}{\left(\dfrac{\partial x}{\partial u}\right)^2 + \left(\dfrac{\partial y}{\partial u}\right)^2} + \lambda y = a, \tag{3.9}$$

where $x = x(u, v)$ and $y = y(u, v)$ are two conjugate harmonic functions regular in the unit strip.

Next, we will linearize the problem. We assume that the free surface deviates little from the straight line $y = 1$ and that the flow is "almost" translatory:

$$x = u + \varphi, \ y = v + \psi,$$

where φ and ψ are small conjugate harmonic functions; then

$$\frac{1}{\left(\frac{\partial x}{\partial u}\right)^2 + \left(\frac{\partial y}{\partial u}\right)^2} = \frac{1}{\left(1 + \frac{\partial \varphi}{\partial u}\right)^2 + \left(\frac{\partial \psi}{\partial u}\right)^2} \approx 1 - 2\frac{\partial \varphi}{\partial u}$$

and (3.9) assumes for $v = 1$ the form

$$1 - 2\frac{\partial \varphi}{\partial u} + \lambda(1 + \psi) = a$$

or, by virtue of the conjugateness of φ and ψ,

$$I_0 = \frac{\partial \psi}{\partial v} - \tfrac{1}{2}\lambda\psi = \tfrac{1}{2}(1 + \lambda - a) = b. \tag{3.10}$$

Our problem has now been reduced to the following: to construct a harmonic function ψ, regular in the strip $0 < v < 1$ and satisfying on its boundary for $v = 1$ the condition (3.10) and on $v = 0$ the flow condition $\psi = 0$.

The solution of this boundary value problem may be obtained in closed form. In fact, consider the harmonic function

$$\psi = C \sinh Av \cdot \cos Au + kv. \tag{3.11}$$

This function is regular in $0 < v < 1$; for $v = 0$, we have $\psi = 0$, and we must still determine the constants C, A and k in such a manner that the conditions (3.10) and $\psi(0, 1) = 0$ are fulfilled. We find

$$CA \cosh A \cos Au - \tfrac{1}{2}C\lambda \sinh A \cos Au + k - \tfrac{1}{2}\lambda k \equiv b,$$

$$C \sinh A + k = 0.$$

These two relations render three equations for the determination of the three constants:

$$k = -C \sinh A, \; k = \frac{b}{1 - \frac{1}{2}\lambda},$$

$$A \cosh A - \tfrac{1}{2}\lambda \sinh A = 0,$$

or

$$k = \frac{b}{1 - \frac{1}{2}\lambda}, \quad C = \frac{-b}{(1 - \frac{1}{2}\lambda) \sinh A},$$

$$A \cosh A - \tfrac{1}{2}\lambda \sinh A = 0. \tag{3.12}$$

The problem has been reduced to the solution of the last of these equations. As also in the general case, the equation (3.12) has no solution for $\lambda < 2$, but for $\lambda \geq 2$ it has the unique solution

$$A = A(\lambda), \qquad \lambda \geq 2. \tag{3.13}$$

The function $A(\lambda)$ vanishes for $\lambda = 2$ and increases for $\lambda > 2$; for larger λ, one can write approximately

$$A \approx \tfrac{1}{2}\lambda \qquad (\lambda \gg 2),$$

for $\lambda \approx 2$,

$$A \approx \sqrt{\tfrac{3}{2}(\lambda - 2)}.$$

If we select as free surface Γ the surface Γ_1, obtained in the linear formulation, and then construct a conformal map $w = f(z, \Gamma_0, \Gamma_1)$ of the region $D(\Gamma_0, \Gamma_1)$ on to the strip $0 < v < 1$, the condition (3.8) will not be strictly fulfilled. However, by use of the variational principles, it is not difficult to obtain estimates of the oscillations of $I(x, \Gamma_1)$ and, under the hypothesis that a solution of the exact problem exists, of the deviation of Γ_1 from the exact surface Γ.

In fact, for sufficiently small values of $\nu = Ac$, the curvature of the curve Γ_1 will be of order ν; however, in that case the deviation of $|f'(z, \Gamma_0, \Gamma)|^{-2}$ from $1 - 2(\partial\varphi/\partial n)$ will be of order ν^2.

Thus,

$$|I(x, \Gamma_1) - a| < k\nu^2,$$

where k is some constant and we have proved

LEMMA 3.1. *An approximate solution corresponding to the linearized problem satisfies the boundary condition* (3.8) *exactly apart from terms of order* ν^2*, where* ν *is the product of the frequency of the wave* A *and its height* c.

We note yet

LEMMA 3.2. *Let* $\lambda > 2$ *and* $A = A(\lambda)$ *be the function determined by* (3.13); *if* Γ *given by* $y = y(x)$ *is a periodic even function with period* $2\pi/A$ *which vanishes at all points* $n\pi/A$ $(n = 0, \pm 1, \pm 2)$, *then the variation of* $I(x, \Gamma)$ *is larger than some constant* k, *multiplied by the variation of* $y(x)$:

$$\operatorname{var} I(x, \Gamma) \geq k \operatorname{var} y(x).$$

The proof of this lemma will not be presented here and only brief remarks of a general nature will be made. The proof is quite simple, if the functional I is replaced by the linear functional (3.10). In the general case, a proof can be obtained by construction of an upper bound which can be derived on the basis of a particular variational principle for the correspondence of boundaries in the case of conformal transformations of strips. One can obtain a quantitative estimate and expression for k in terms of λ by use of the formula for the variation of the derivative in terms of the variation of the boundary of the transformed region (cf. Chapter I).

It is not difficult to derive from Lemmas 3.1 and 3.2 an estimate of the accuracy of an approximate solution.

THEOREM 3.1. *If a solution of the functional equation* (3.8) *exists, the curve* $\Gamma : y = y(x)$ *has the period* $2\pi/A$ [*where* $A = A(\lambda)$ *is determined by* (3.13)] *and one has a small oscillation* $\delta = \operatorname{var} y(x)$, *then the deviation of this solution from the linear solution which corresponds to the same parameters has the order* δ^2.

In fact, by Lemma 3.1, the approximate solution satisfies the equation (3.8) exactly apart from quantities of order δ^2; on the other hand, by Lemma 3.2, in the case of variations of the approximate waves by δy, the variation of $I(x, \Gamma)$ is of the same order. Hence, if the correct solution exists, it can deviate from the approximate one only by a quantity of order δ^2.

3.6. Rayleigh waves

The linear theory studied above does not account for many important facts, revealed first by experiment, for example, that in the case of an increase of the height of a wave the crest becomes steeper than the trough, that the length of a wave grows with its height (for fixed depth and velocity of propagation), etc. The linear theory and related theories also cannot explain the so-called solitary wave. The approximate non-linear theory of shallow waves, including the theory of solitary waves (the curve Γ has a single maximum for $|x| < \infty$) was first developed by Rayleigh. We will study now Rayleigh waves on the basis of the approximate expressions for conformal transformations of narrow strips.

We replace in the boundary condition (3.7) $|f'|$ by its expression (2.16). Omitting the remainder term R, we obtain

$$\left(\frac{H}{y}\right)^2 (1 + \tfrac{1}{3}yy'' + \tfrac{1}{3}y'^2)^2 + 2gy = C$$

or, to the same degree of accuracy,

$$\left(\frac{H}{y}\right)^2 (1 + \tfrac{2}{3}yy'' + \tfrac{2}{3}y'^2) + 2gy = C.$$

After simplifications, the last relation assumes the form

$$y'' = -\frac{3}{2}\,\frac{1 + \frac{2}{3}y'^2}{y} + \alpha y - \beta y^2, \tag{3.14}$$

where α and β are two supplementary constants determined by the physical parameters of a problem. The differential equation (3.14) can be integrated in closed form; if we assume for this purpose, without restricting generality, that

$$y'(0) = 0, \qquad y''(0) < 0$$

(for $x = 0$, we have the crest of the wave), then we find as a result of the integration a system of waves which depends on three parameters: α, β and a constant of integration.

In order not to overburden the study with computations, we will still simplify the equation (3.14) by neglecting the term in y'^2. This step is justified if y'^2 is small compared with the term yy''. We will show that this is actually so for a sufficiently wide class of cases which includes that of the solitary wave.

Consider the equation

$$y'' = -\frac{3}{2}\frac{1}{y} + \alpha y - \beta y^2 \equiv \varphi(y), \tag{3.15}$$

obtained in the case of the simplification proposed above; first of all, we note that the equation (3.15) admits a bounded solution (on the entire axis) only if

$$-\frac{3}{2}\frac{1}{y} + \alpha y - \beta y^2 = 0$$

has positive roots, which we will denote by y_0 and y_1. In that case the graph of $\varphi(y)$ has the form shown in fig. 13.

In the case of these limitations on the parameters α and β, we construct a system of integral curves which start from different points of the y-axis in horizontal directions. For $y(0) = y_1$, one will have a straight line, for $y(0) = y_1 + \delta$ with δ sufficiently small, we will have a periodic curve with steep crests and less steep troughs, where for $\delta \to 0$ the period tends to $l = 2\pi/[\varphi'(y_1)]^{\frac{1}{2}}$ (fig. 14). With increasing δ, the minimum of $y(x)$ will decrease and the period $\omega = \omega(\delta)$ increase, where

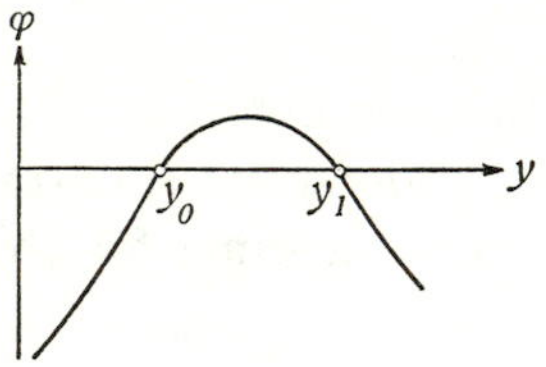

Fig. 13

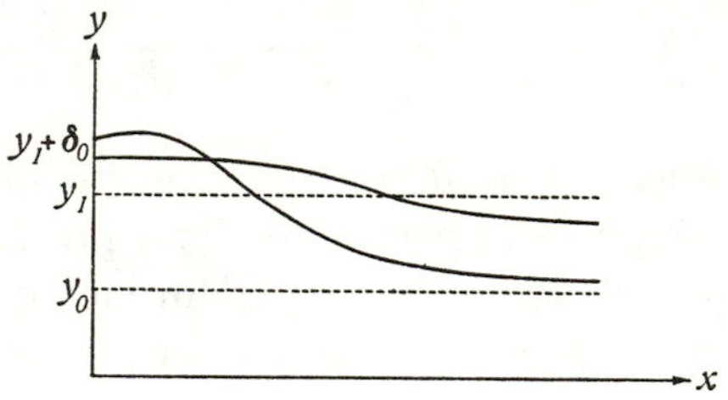

Fig. 14

for $\delta \to \delta_0$, with δ_0 a constant expressible in terms of the basic parameters α and β, the period grows beyond all bounds and our periodic function becomes a curve with a single maximum at the point $x = 0$ (fig. 14). The last curve can be considered to be an approximation to the solitary wave.

Instead of the parameters α and β, one can select other parameters with physical significance. For example, one can use the "depth" y_1 and the length of the linear wave $l = 2\pi/\lambda$ where $\lambda = \sqrt{\varphi'(y_1)}$. We can compute α and β in terms of y_1 and l from the equations

$$-\frac{3}{2}\frac{1}{y_1} + \alpha y_1 - \beta y_1^2 = 0, \ \lambda^2 = \frac{4\pi}{l^2} = \frac{3}{2}\frac{1}{y_1^2} + \alpha - 2\beta y_1$$

which render

$$\alpha = \frac{9}{2}\frac{1}{y_1^2} - \frac{4\pi^2}{l^2}, \ \beta = \frac{3}{y_1^3} - \frac{4\pi^2}{y_1 l^2}. \tag{3.16}$$

Other possible parameters are the root y_0 and the height of the solitary wave U. Obviously, the root y_0 is equal to the depth of the channel when a solitary wave of height U is propagated along it. If we are to seek the integral curves of the equation (3.15) for these values of y_0 and U with $y(0) = y_1 + \delta$, $0 < \delta < U - y_1$, we will obtain a system of waves for which, however, the depth of the basin will be changed; in fact, for a decrease δ from $U - y_0$ to 0, the depth of the channel changes from y_0 to y_1.

From the point of view of mechanics, the most natural basic parameters are the mean depth

$$h = \lim_{L\to\infty} \frac{1}{L} \int_{-L}^{+L} y(x)\, dx \tag{3.17}$$

and the amount of flow H. For fixed values of h and H, we can construct a system of all possible waves; this will be a one-parameter system. As before, we can select as the parameter the height of the crest of the wave $y(0) = y_1 + \delta$. For $y_1 + \delta = U$, we obtain the solitary wave; for a smaller δ, we must increase y_0 in order to maintain the value of h, and for $y\ (0) = y_0 \to y_1$, we obtain the linear wave.

In order to establish a perfect foundation for an approximate theory, it is sufficient to establish the following facts:

1) there exist exact solutions corresponding to any values of the parameters h, H and δ which satisfy the relations

$$h_0 \leq h \leq h_1, \; H_0 \leq H \leq H_1, \; \delta_0 \leq \delta \leq \delta_1;$$

2) the solution $\gamma : y = y(x, h, H, \delta)$ is stable, i.e., if one has another solution $\tilde{\gamma} : y = \tilde{y}(x)$, $\tilde{y}(0) = y_1 + \delta$, such that

$$\max |I(\tilde{\gamma}, x) - a| < \varepsilon,$$

then

$$|\tilde{y}(x) - y(x, h, H, \delta)| < \nu(\varepsilon),$$

where $\nu(\varepsilon)$ is a universal function, $\lim_{\varepsilon \to 0} \nu(\varepsilon) = 0$;

3) if the approximate solution γ_0 is substituted into the functional I, then

$$|I(\gamma_0, x) - a| < \varepsilon', \tag{3.18}$$

where ε' is such that $\nu(\varepsilon')$ is small compared with the amplitude of the waves $[\nu(\varepsilon') \ll y_1 + \delta - h]$.

In the section devoted to exact solutions, we will realize this scheme for the linear theory. We will note now an estimate for ε' in the inequality (3.18), and, in particular, solve the problem regarding the values of the parameters for which the quantity ε' will be essentially smaller than the principal terms in the approximate expression for the functional $I(\gamma, x)$.

Estimate of ε'. It will be sufficient for the solution of the problem posed to obtain estimates for the first four derivatives of the integral of the equation (3.15) and of the remainder term R.

In order to simplify the manipulations, we will impose beforehand additional limitations on the parameters. We will use as basic small quantity the value $y = \eta$ for which the function $\varphi(y)$ attains a maximum; we will assume this maximum to be equal to $k\eta^\nu$, where k is some fixed constant and ν a positive constant smaller than unity. This restriction corresponds to the fact that the height of the corresponding solitary wave is small compared with the depth of the channel. In this case, within the limits of the accuracy adopted, we can take instead of (3.15) the equation

$$y'' = A\eta^{\nu} - \frac{B}{\eta^3}(y - \eta)^2 = \tilde{\varphi}(y), \tag{3.19}$$

where A and B are certain fixed constants. The first integral of (3.19) has the form

$$\left(\frac{dy}{dx}\right)^2 = 2A\eta^{\nu}y - \tfrac{2}{3}B\left(\frac{y-\eta}{\eta}\right)^3 + C, \tag{3.20}$$

where the integration constant C is determined by the condition that for $x = 0$ one has $y = \eta + \delta$ and $y' = 0$, whence

$$C = \tfrac{2}{3}B\left(\frac{\delta}{\eta}\right)^3 - 2A\eta^{\nu}(\eta + \delta). \tag{3.21}$$

Using (3.20) and (3.21), an estimate of $|y'|$ may be obtained. For this purpose we will find, first of all, the roots y_0 and y_1 of the equation $\tilde{\varphi} = 0$:

$$y_{0,1} = \eta \pm \sqrt{\frac{A}{B}}\,\eta^{(3+\nu)/2}. \tag{3.22}$$

We conclude from (3.22) that the maximum value of δ (corresponding to the solitary wave) is of order $\eta^{(3+\nu)/2}$. Substituting into (3.20) and (3.21) the maximum value of δ and replacing y by its value $y_1 = \eta + \delta_1$ corresponding to the maximum of y', we obtain

$$|y'| \leq M\left(\frac{\delta}{\eta}\right)^{3/2} = M_1\eta^{3/4(1+\nu)}.$$

In addition, one has directly from (3.19) the inequality

$$|y''| \leq M_2\eta^{\nu},$$

and, after differentiating $\varphi(y)$ with respect to x, the estimates required for y''' and $y^{\prime v}$:

$$y''' = \frac{-2B}{\eta^3}(y - \eta)y',$$

$$y^{\prime v} = -\frac{2B}{\eta^3}y'^2 - \frac{2B}{\eta^3}(y - \eta)y'';$$

using the estimates for y' and y'', we obtain, finally, the upper bounds for these expressions:

$$\begin{aligned} |y'''| &\leq M_3\eta^{-3/4+5/4\nu}, \\ |y^{\prime v}| &\leq M_4\eta^{3/2(\nu-1)}, \end{aligned} \tag{3.23}$$

where M is a fixed constant (expressible in terms of A and B).

Next, consider the estimate for the remainder term. We have

$$\begin{aligned} |R_3| &\leq \tfrac{1}{3}y'^2 + |y'''|\, h^2 \leq N_3\eta^{5/4(1+\nu)}, \\ |R_4| &\leq \tfrac{1}{3}y'^2 + |y^{\prime v}|\, h^3 \leq N_4\eta^{3/2(1+\nu)}, \end{aligned} \tag{3.24}$$

where N is some constant.

It is seen from these estimates that for any values of ν, $0 \leq \nu \leq 1$, the omitted terms are small compared with the retained principal terms, whence it is also concluded that the omission of the term $y'^2/3$ does not reduce the order of the accuracy. These estimates are sufficient for the exact theory in the case when the length of the waves has the order of the length of the linear waves, i.e., $1/\sqrt{|\varphi'(y_1)|}$. For the exact theory of long waves and, in particular, of the solitary wave, it is important to have estimates of the accuracy which take into account small curvatures and the deviations of the troughs. This part of the wave corresponds to values of y which are close to y_0; hence we conclude that in this case one may replace the function $\varphi(y)$ by its principal linear part near the point $y = y_0$. We arrive thus at the differential equation

$$y'' = \varphi'(y_0)(y - y_0) \tag{3.25}$$

and obtain for the trough of the wave

$$y = y_0 + Ce^{-\lambda x},$$

where

$$\lambda = \frac{1}{\sqrt{|\varphi'(y_0)|}}.$$

Hence, for values of x which are smaller than the half-period $l/2$, but large compared with λ, the derivatives y', y'', y''', $y^{\prime v}$ will be of order $e^{-\lambda x}$; combining this result with the estimates derived earlier for the same derivatives, we obtain

$$\begin{aligned} |y'| &\leq M\eta^{3/4(1+\nu)}e^{-m\lambda x}, \\ |y''| &\leq M\eta^{\nu}e^{-m\lambda x}, \\ |y'''| &\leq M\eta^{-3/4+5/4\nu}e^{-m\lambda x}, \\ |y^{\prime v}| &\leq M\eta^{3/2(\nu-1)}e^{-m\lambda x}. \end{aligned} \tag{3.26}$$

It follows from (3.26) that the wave shape satisfies conditions for which (2.16) is true with a remainder term the order of which is higher than that of the terms included.

The channel of variable depth. The present approximate theory can be extended to the case of steady motion of a fluid in a channel of variable depth. The problem is the following: Given the curve $\Gamma_0 : y = y_0(x)$, $|x| < \infty$, where $y_0(x)$ is a twice differentiable bounded function, it is required to construct a curve $\Gamma : y = y(x)$, $y(x) > y_0(x)$ such that one has along Γ

$$I(x, \Gamma) = |f'(z, \Gamma_0, \Gamma)|^2 + \lambda y = C. \tag{3.27}$$

It has been proved above that for certain additional restrictions referring to the bottom Γ_0 for fixed C and for sufficiently small λ the solution of this problem exists and is unique, where the width H of the strip $0 < v < H$ of the w-plane (amount of fluid flow) will be completely determined by the bottom Γ_0 and the constants C and λ.

If the functional $I(\Gamma)$ is replaced in an approximate manner by the differential operator of (2.16), the problem of the determination of Γ reduces to the integration of an ordinary second order differential equation. A formal solution of such an equation always exists and depends on two arbitrary parameters. The arbitrariness is reduced, if we require the solution to be close to the solution of the basic boundary value problem; in this case, several conditions must be fulfilled:

1) the integral curve $y = y(x)$ must lie above the bottom $y = y_0(x)$, i.e., the inequality $y(x) > y_0(x)$ must apply everywhere;

2) the function $y(x)$ must be bounded;

3) the strip $y_0(x) < y < y(x)$ must satisfy conditions under which the remainder term R in (2.16) will be small compared with the other terms.

As in the case of the exact theory, there exists for λ sufficiently small a solution of the equation obtained which satisfies the first two conditions and which is unique apart only from a single parameter; a second parameter can only appear starting with a definite value of λ. We will not continue the study of this problem in its general form, but proceed to one particular case where all three conditions happen to be satisfied. This particular case will be required for the exact theory of Rayleigh waves.

The periodic, almost even bottom. In the notation, adopted for the approximate theory of Rayleigh waves, we will assume that $\eta = h$ is the basic small quantity and impose on $y = y_0(x)$ the following conditions:

1°. $y_0(x)$ is a periodic function with half-period $l \gg \sqrt{\eta}$:

$$y_0(x + 2l) = y_0(x);$$

for $x = 0$, $y_0(x)$ attains its single maximum in the interval $|x| < l$;

2°. Γ_0 belongs to the strip $kh^2 \leq y \leq (k + k_1 h)h^2$, where k_1 is a finite quantity:

$$kh^2 \leq y_0(x) \leq (k + k_1 h)h^2;$$

3°. Γ_0 has a small gradient:

$$|y_0'(x)| \leq \nu h^2$$

where ν is small together with h. We will select for the parameters λ and C the values which we had adopted earlier for the study of the approximate theory of the solitary wave (assuming $\nu = 1$).

We will now construct the differential equation for an approximate solution of the problem of the motion of a heavy fluid in a channel. Let the flow of fluid be equal to H, then one can write for $|f'(z, \Gamma_0, \Gamma)|$ at an arbitrary point of Γ, using (2.16),

$$|f'(z, \Gamma_0, \Gamma)| = \frac{H}{y(x) - y_0(x)} (1 + \tfrac{1}{3}yy'' + R). \qquad (3.28)$$

It is readily shown that, if the unknown curve $y = y(x)$ satisfies the conditions accepted above for the study of the theory of Rayleigh waves,

$$\begin{gathered} h - h^2 \leq y(x) \leq h + 2h^2, \ |y'(x)| < kh^{\frac{3}{2}}, \ |y''(x)| < kh, \\ |y'''(x)| < k\sqrt{h}, \end{gathered} \qquad (3.29)$$

then the remainder term R is of higher order than h^2:

$$|R| < k(\nu + h^{\frac{1}{2}})h^2. \qquad (3.30)$$

In fact, if $y_0(x)$ is assumed to be a constant, which is equal to the value of $y_0(x)$ at the point under consideration, in accordance with the preceding results the remainder term R will be of order $h^{\frac{5}{2}}$; there remains still to obtain an estimate of the variation of $|f'|$ when $y_0 =$ const. is replaced by a majorant curve consisting of segments of straight lines with gradients $\pm\ \nu h^2$ and parallel to the x-axis. Using the variational principle relating to boundaries, one can derive for the variation of $|f'|$ the estimate

$$\delta |f'| < k \int_0^\infty \frac{1}{(h + x)^2} e^{-(x/h)} \delta y \, dx = k \int_0^\infty \frac{\nu h^2 e^{-(x/h)} x \, dx}{(h + x)^2} =$$

$$= k\nu h^2 \int_0^\infty \frac{\tau}{(1 + \tau)^2} e^{-\tau} d\tau = k_1 \nu h^2.$$

Thus, with the proposed degree of approximation, the solution of the problem posed leads to the differential equation

$$\frac{H^2}{(y - y_0)^2}(1 + \tfrac{2}{3}yy'') + \lambda y = c$$

which assumes, after some simple transformations, the form

$$y'' = -\frac{3}{2}\frac{1}{y} + \frac{3y}{2H^2}(c - \lambda y)\left(1 - \frac{y_0}{y}\right)^2. \tag{3.31}$$

Generalizations. The above non-linear approximate theory of steady motion of a heavy fluid has been developed considerably during the last year (Moiseev, Bichinkov). Equations of the type (3.31) have been studied in detail for different profiles of bottoms and the case has been investigated when a hydrofoil is located below the surface of the heavy fluid.

I will make here only one interesting remark relating to the uniqueness of flows with depth h and fluid transport H given at $-\infty$. Consider the motion in the case when h is the smaller of the roots of the equation $y(x) = 0$. Then, by the above results, for given depth h and fluid transport H (at the point $-\infty$) a steady motion is unique in the class of curves close to the straight line $y = h$. However, if we do not restrict ourselves to curves close to $y = h$, we obtain together with the true solution $y = h$ a solution $y = y(x)$, corresponding to the solitary wave.

Such multiplicity is naturally also obtained in the case of a curvilinear bottom, i.e., the presence of a small trench or mound on the bottom may yield two (an infinite set of) steady flows with a given depth and fluid transport.

3.7. The exact theory

We will show now how we may construct with the help of the variational principle exact solutions of the corresponding problems, based on the approximate solutions. We will apply the method in equal measure to the construction of an exact solution, close to the solution of the linear theory, and of one close to the non-linear Rayleigh solution; in particular, a proof of the existence of the solitary wave has been obtained first by this method.

We will now give the basic steps for waves close to linear waves. We will prove the existence of a wave close to the linear wave considered above (cf. § 3.5). First of all, we will fix the length of the wave (a fixed finite quantity) and its height $\mu = \text{var}\, y(x)$ which we will assume to be sufficiently small.

Let $\Gamma : y = y(x)$ denote the curve corresponding to the linear theory. By Theorem 3.1, if a solution in the exact formulation exists, it deviates from Y by a quantity which is smaller than $k\mu^2$. Consequently, the unknown solution must belong to the class of curves $\{y\}$ located in the strip

$$Y(x) - k\mu^2 < y < Y(x) + k\mu^2.$$

We will now map the strip $D(\Gamma_0, \gamma)$, where Γ_0 is the x-axis, on to the unit strip $0 < v < 1$ and denote by γ_m the stream line corresponding to the straight line $v = m/n$, $m = 1, 2, \cdots, (n-1)$, where n is some fixed integer.

In addition, we note two essential properties of the curve γ_m which follow from the maximum principle.

1°. If we denote by $\Gamma_m : y = Y_m(x)$ a stream line corresponding to an approximate wave, then the manifold $\{\gamma_m\}$ will belong to the strip

$$Y_m(x) - k_1 \frac{m}{n} \mu^2 < y < Y_m(x) + k_1 \frac{m}{n} \mu^2, \qquad (3.32)$$

where k_1 is close to k and does not depend on n.

2°. It follows from the single-valuedness of the function $y(x)$ of the manifold $\{y\}$ that the gradients and curvatures of all curves γ_m will be bounded, where, if $m/n \leq k < 1$, the curvature $k_m(x)$ and gradient $y_m'(x)$ of the curve γ_m will be bounded quantities depending only on k (for fixed μ):

$$|k_m(x)| < \alpha(k), \quad |y_m'(x)| < \beta(k), \qquad (3.33)$$

where for $k \to 0$ we have $\alpha \to 0$ and $\beta \to 0$.

Noting this we denote by $\tilde{\gamma}_m : y = \tilde{y}_m(x)$ the manifold of all curves satisfying the conditions (3.32) and (3.33). We will prove the existence of the curve γ, satisfying the condition (3.27):

$$I(x, \gamma) = |f_m'(z, \tilde{\gamma}_m, \gamma)|^2 + \lambda y = a,$$

where f_m, $m = m_0, m_0 - 1, \cdots, 1, 0$ are functions mapping the strip $\tilde{y}_m(x) < y < y(x)$ conformally on to the strip $m/n < v < 1$. In order to construct the unknown curve γ, it will be shown that it will deviate from the linear wave Γ by a quantity of order ν^2 and that $y(x)$ will have the same period as $Y(x)$.

The proof has two parts:

I. By virtue of considerations analogous to those above, k_0 can be found such that for $m/n \geq k_0$ there exists a curve γ satisfying the condition (3.27) and the stated supplementary conditions. In fact, for k_0 sufficiently close to unity, if we were to vary the curve γ (close to the linear wave), the variation of $|f_m'|$ will be of order $\delta y/(1 - k_0)$, while the variation of λy will be equal to $\lambda \delta y$. Thus, the term $\lambda \delta y$ is not essential and we find ourselves in a region of applicability of the basic variational principle. Now denote by m_0 the smallest number for which $m/n \geq k_0$. By (I), there exists for any curve of the class $\{\tilde{\gamma}_m\}$ a corresponding $\tilde{\Gamma}_m$; besides, by Lemma 3.2, our solution is stable: for variations of the curve $\tilde{\gamma}_{m_0}$, an upper bound of the variations of $\tilde{Y}_{m_0}$ is given by

$$|\delta \tilde{Y}_{m_0}| < k \max |\delta \tilde{y}_{m_0}(x)|, \tag{3.34}$$

where k is a constant which depends only on the selected parameters of the problem and on the amplitude of the wave μ fixed earlier. We note still that (3.34) also applies for any $m < m_0$, where the constant k may be given the same value for all m independently of n. In fixing k in this manner, we will assume that n is large compared with k, i.e., $n \gg k$.

Starting with the above class of solutions $\tilde{\Gamma}_{m_0}$, we will construct successively solutions $\tilde{\Gamma}_{m_0-1}, \tilde{\Gamma}_{m_0-2}, \cdots, \tilde{\Gamma}_0$. We assume that the existence of $\tilde{\Gamma}_m$ has been proved for any $\tilde{\gamma}_m$ out of the class $\{\tilde{\gamma}_m\}$ selected earlier; then we take an arbitrary curve $\tilde{\gamma}_{m-1}$ from the class $\{\tilde{\gamma}_{m-1}\}$. Among the curves of the class $\{\tilde{\gamma}_m\}$ consider a subclass $\{\tilde{\tilde{\gamma}}_m\}$ the curves of which lie below the curve $\tilde{\gamma}_{m-1}$ and satisfy, in addition, the conditions

$$|\tilde{\tilde{y}}'_m| \leq k', \ |\tilde{\tilde{y}}''_m| < k'',$$

where k' and k'' must be given sufficiently large values. By induction, a wave $\tilde{\Gamma}_m$ corresponds to every curve $\tilde{\tilde{\gamma}}_m$; on the other hand, we may map the strip $\tilde{y}_{m-1}(x) \leq y \leq \tilde{\tilde{y}}_m(x)$ conformally on to the strip $(m-1)/n < v < m/n$. The curve $\tilde{\Gamma}_m$ can be considered as the wave [solution of the equation (3.27)] corresponding to $\tilde{\gamma}_{m-1}$, if in the case of conformal mappings of the strips $\tilde{\tilde{y}}_m < y < Y_m$ and $\tilde{y}_{m-1} < y < \tilde{\tilde{y}}_m$ on to the strips $m/n < v < 1$ and $(m-1)/n < v < m/n$ the extensions of both mappings along the curve $\tilde{\tilde{\gamma}}_m$ will be the same. The problem reduces thus to that of patching, i.e., to a proof of the existence of a curve $\tilde{\tilde{\gamma}}_m$ for which

$$I(x, \tilde{\tilde{\gamma}}) = \log |f'(z, \tilde{\tilde{\gamma}}_m, \tilde{\Gamma}_m)| - \log |f'(z, \tilde{\gamma}_{m-1}, \tilde{\tilde{\gamma}}_m| = 0,$$
$$z \in \tilde{\tilde{\gamma}}_m.$$

It is not difficult to show that we arrive by a choice of sufficiently large values of k', k'' and n at conditions which are sufficient for the applicability of the basic principle: if for some curve $\tilde{\tilde{\gamma}}_m$ of our class $I > 0$, then $\tilde{\tilde{\gamma}}_m$ can always be deformed in such a manner that the new curve will belong to the same class and that along this curve max I is decreased. This theorem proves completely the existence. We have thus constructed a solution close to the linear solution.

In order to prove the theorem of the existence of a wave, close to a linear wave, we have employed the following two essential facts:

1) the existence of a solution which satisfies the basic functional equation exactly apart from second order quantities which are small compared with the height of the wave,

2) the stability of the wave.

These same facts hold true if we select in place of the linear wave a non-linear wave an approximate theory of which has been studied in detail above. The scheme investigated leads us to the existence and uniqueness of systems of non-linear waves (including the solitary wave) for the values of the parameters used in the approximate theory. We have limited ourselves here to waves which are close to the linear waves, because a detailed study of the non-linear case requires technical details which are, in principle, straight forward, but cumbersome.

Approximate methods, machine computations and exact solutions. From the above method of proof of the existence theorem and from the constructive procedure on which it is based one can obtain different approximate methods of solution of corresponding problems of wave theory without the assumption of small amplitudes.

1. In accordance with the linear theory, construct a wave with small amplitude. Then map the region occupied by the wave conformally on to a strip and proceed to a wave with large amplitude which does not differ greatly from the constructed wave. For this purpose, one can seek the new wave again in the linear formulation, etc.

2. Subdivide the region occupied by the fluid by means of stream lines into n parts corresponding to the strips $m/n < v < (m+1)/n$. The derivatives at the boundary of the functions mapping the strip $D(\gamma_m, \gamma_{m+1})$ conformally on to the strip $m/n < v < (m+1)/n$ can be computed by the approximate formula for the mapping of narrow strips. The problem of the construction of waves is thus reduced to the solution of systems of second order differential equations; one equation, corresponding to the free surface, will have the form (3.27), while the remaining equations will express the "patching" conditions of the strips.

In conclusion, we will note the possibility of the construction of a complete theory of stationary waves for all depths and amplitudes ranging from linear to solitary (for small depths) and from linear to Stokes waves (for deep fluids). The establishment of such a universal theory can be achieved by the following scheme:

1°. Using one of the approximate methods (employing modern electronic computers), construct approximate solutions over a large range of parameters. The necessary density of this "grid" of solutions can be assessed either beforehand or after computation of a set of solutions as a first approximation:

2°. By the method of determining upper bounds for the solutions obtained, one can find (again employing digital computers) sufficiently exactly the characteristics of these solutions which are necessary for the applicability of the variational method; these characteristics are the smoothness of the solution, the first four derivatives along the boundaries, the deviation of the functional $I(\gamma, x)$ from zero and its variations.

3°. The variational method studied above proves that there exists in the neighbourhood of every approximate solution a manifold of exact solutions which fill the "intervals" between the elements of the system of approximate solutions.

3.8. Generalizations

In concluding the study of the variational methods and the theory of steady wave motion of a heavy liquid, we will still dwell on several problems some of which have been solved, others not, but all of which lie within the same circle of ideas.

We will consider the steady motion of a heavy liquid over an uneven bottom. In the theory of shallow fluids, we have noted the possibility of two independent solutions; we will now consider the formulation of problems with multi-valuedness of a different physical character.

3.8.1. Motion of a fluid over a submarine trench. Let the bottom of the fluid be everywhere plane, except for a segment AB where a trench occurs. Under these conditions the motion of a heavy fluid is subject to the following conditions:

1) for $x = \pm\infty$, the depth h of the fluid is given,

2) the amount of fluid flow H is given.

Then, on the basis of the considerations studied above, the motion exists and is unique for H/h sufficiently large. However, the actual flow will be close to another flow of the following type.

We join the points A and B (the boundary of the trench) by a curve γ and consider two flows:

a) inside the region bounded by the arc AB of the trench and the arc γ, the flow of the liquid with constant vorticity ω is described by the equations

$$\frac{\partial u}{\partial x} + \frac{\partial v}{\partial y} = 0, \qquad \frac{\partial u}{\partial y} - \frac{\partial v}{\partial x} = \omega; \tag{3.35}$$

b) the flow of the liquid, for the same conditions at infinity as in the basic flow, streaming along the bottom of the channel past the points A and B and the arc γ (fig. 15). The arc γ must be selected in such a manner that the velocities of the two flows match. In the arc of such a "secondary" regime in the trench, we will have on the free surface of the flow instead of a cavity a hump.

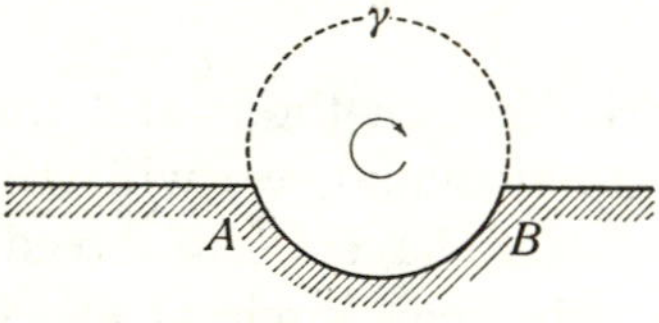

Fig. 15

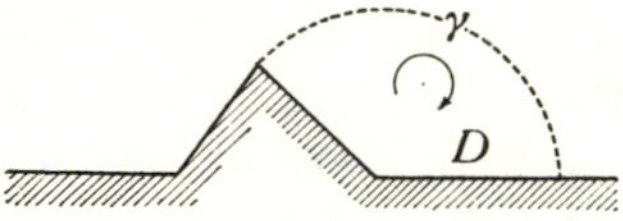

Fig. 16

3.8.2. Motion over a bottom with a ridge. In the above formulation, we assume that the plane bottom of the channel has a hump instead of a trench (fig. 16). Using a curve γ, we join the vertex of the ridge to the bottom of the channel in the direction of the flow and consider again two flows:

a) a flow with constant vorticity [described by the system (3.35)] in the region D bounded by part of the bottom of the channel and the curve γ;

b) the potential flow of the heavy liquid in a region obtained by augmenting the original region by the region D.

The curve γ and the quantity ω must be selected such that the velocities of both flows along it match. In such a secondary regime, a hump will occur above the trench instead of a depression.

3.8.3. Spillway with singularities. In recent times, the problem of the spillway has been the subject of many investigations aiming at more precise results. Let the section Γ of the bottom of a spillway consist of the half straight line $x = 0$, $-\infty < x < 0$, the transition $y = y(x)$, $y'(x) < 0$ and the half straight line $y = -1$, $1 < x < \infty$ (fig. 17). In this case we are dealing with a flow of a heavy liquid subject to the conditions:

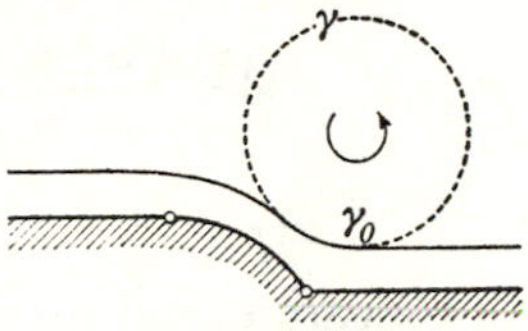

Fig. 17

1) the depth h of the fluid is given,

2) the amount of fluid flow H is given.

Under these conditions, one may consider a motion of a fluid which is essentially different from the ordinary flow with singularities. We consider a closed curve γ which has with the free surface of the flow an arc γ_0 in common and construct two flows:

a) the flow inside γ with constant vorticity,

b) the basic flow with a free surface occurring between the bottom of the spillway and γ, where the free surface has with γ an arc γ_0 in common. One has to find the curve γ and a value of ω such that

1°. the velocity of the flow inside γ along γ_0 and the basic flow match,

2°. the velocity of the flow on γ and on the free surface of the basic flow satisfy the condition

$$I(\gamma, x) = V^2 + \lambda y = C. \tag{3.36}$$

CHAPTER IV

QUASI-CONFORMAL MAPPINGS

The variational method developed above for the solution of boundary value problems for the Laplace equation is based on a small number of properties of conformal mappings. Among these the principal ones are Riemann's theorem on the existence of conformal transformations and the basic internal and boundary principles of such mappings, i.e., Lindelöf's and Montel's principles. Thus, the variational method may be extended to those systems of partial differential equations for which the properties of conformal mappings are valid and the application of a variational principle is guaranteed.

In this chapter, we will describe the class of systems of partial differential equations and their related mappings (which will be said to be quasi-conformal) to which the variational principles of conformal transformations may actually be extended. Such a geometrisation of the theory of systems of partial differential equations will be of great interest. In addition, within the framework of this theory, one can solve for the first time a number of problems of steady motion of gases in plane cases as well as in cases of axial symmetry.

4.1. The concept of the quasi-conformal map

Let there be given a system of partial differential equations

$$
\begin{aligned}
F_1\left(x, y, u, v, \frac{\partial u}{\partial x}, \frac{\partial u}{\partial y}, \frac{\partial v}{\partial x}, \frac{\partial v}{\partial y}\right) &= 0, \\
F_2\left(x, y, u, v, \frac{\partial u}{\partial x}, \frac{\partial u}{\partial y}, \frac{\partial v}{\partial x}, \frac{\partial v}{\partial y}\right) &= 0.
\end{aligned}
\qquad (4.1)
$$

We will say that a homomorphic transformation of a region D on to a region Δ

$$w = f(z) = u(x, y) + iv(x, y) \tag{4.2}$$

is a quasi-conformal transformation of D on to Δ, corresponding to the system (4.1), if the functions u and v constituting (4.2) satisfy this system.

The simplest classical problem which reduces to quasi-conformal mappings is the problem of the conformal mapping of a surface on to a plane region. In fact, let it be required to transform a section of the surface S, given in Cartesian co-ordinates (x, y, s) by the relations

$$s = \Phi(x, y), \quad x^2 + y^2 < 1 \tag{4.3}$$

on to the circle $|w| < 1$ of the plane $w = u + iv$. This means that we have to establish a correspondence between the points of the surface (4.3) and the points of the circle $|w| < 1$, such that any infinitesimal circle on the surface becomes an infinitesimal circle in the w-plane.

It is readily seen that this problem of the conformal mapping of a surface is equivalent to the problem of some quasi-conformal transformation of the circle $|z| < 1$ of the $z = x + iy$ plane on to the circle $|w| < 1$. In fact, every infinitesimal circle on S projects into an ellipse $\mathscr{E}$ in the z-plane. The characteristics of the ellipse $\mathscr{E}$, i.e., the direction θ of its major axis and the ratio p of the major to the minor axes are easily expressed in terms of grad Φ; thus, one has

$$\begin{aligned} \theta(z) &= \arg \operatorname{grad} \Phi(x, y) + \frac{\pi}{2}, \\ p(z) &= \frac{1}{\cos \varphi}, \quad \text{where } \tan \varphi = |\operatorname{grad} \Phi(x, y)|. \end{aligned} \tag{4.4}$$

In order to obtain a conformal mapping of S on to a circle, one only has to map the projection of S, i.e., the circle $|z| < 1$, on to the circle $|w| < 1$ in such a manner that at every point z an infinitesimal ellipse $\mathscr{E}$ with characteristics $p(z)$, $\theta(z)$ becomes an infinitesimal circle of the w-plane.

This geometric characteristic of a quasi-conformal transformation can be expressed in terms of a system of equations of the type (4.1). For this purpose one has only to express the geometric condition in terms of the coefficients of the principal linear part at the point z_0 of the unknown transformation of the unit circles $|z| < 1$ and $|w| < 1$:

$$\begin{aligned} u - u_0 &= \frac{\partial u}{\partial x}(x - x_0) + \frac{\partial u}{\partial y}(y - y_0), \\ v - v_0 &= \frac{\partial v}{\partial x}(x - x_0) + \frac{\partial v}{\partial y}(y - y_0), \end{aligned} \tag{4.5}$$

where the partial derivatives refer to the point $z_0 = x_0 + iy_0$.

In fact, for the transformations (4.2) to map an infinitesimal ellipse $\mathscr{E}$ with characteristics $p(z)$ and $\theta(z)$ into an infinitesimal circle, it is sufficient and necessary that one has between the coefficients the two linear relations

$$\begin{aligned} \frac{\partial u}{\partial x} &= \beta \frac{\partial v}{\partial x} + \gamma \frac{\partial v}{\partial y}, \\ -\frac{\partial u}{\partial y} &= \alpha \frac{\partial v}{\partial x} + \beta \frac{\partial v}{\partial y}, \end{aligned} \tag{4.6}$$

where the coefficients α, β, γ can be expressed in a unique manner in terms of p and θ:

$$\begin{aligned} \alpha &= p \cos^2 \theta + \frac{1}{p} \sin^2 \theta, \\ \beta &= \left(p - \frac{1}{p}\right) \sin \theta \cos \theta, \quad \alpha\gamma - \beta^2 \equiv 1. \\ \gamma &= p \sin^2 \theta + \frac{1}{p} \cos^2 \theta. \end{aligned} \tag{4.7}$$

Thus, the problem of conformal mapping of a surface S on to the circle $|w| < 1$ is equivalent to the problem of the conformal transformation corresponding to a linear system.

Another classical problem which reduces to a quasi-conformal transformation is the problem of the construction of the plane steady gas flow around a given contour. In fact, in the case under consideration, the velocity potential u and the stream function v satisfy the system of equations [1]

$$\frac{\partial u}{\partial x} = k(\lambda)\frac{\partial v}{\partial y},$$
$$\frac{\partial u}{\partial y} = -k(\lambda)\frac{\partial v}{\partial x}, \tag{4.8}$$

where $k(\lambda)$ is a given function of the magnitude of the flow velocity $\lambda = |\text{grad}\, u|$.

Since the contour in the flow must be a stream line, it follows that the stream function v must maintain a constant value on this contour. Therefore the problem of the construction of the gas flow, as in the case of an incompressible fluid, reduces to the transformation of the flow region on to a region bounded by straight lines, parallel to the u-axis. In particular, the motion of gas in a strip $D(\Gamma_0, \Gamma)$ will be determined by a quasi-conformal mapping of D on to a rectilinear strip $0 < v < h$.

As in (4.6), the system (4.8) has a simple geometric interpretation: the quasi-conformal mapping, corresponding to the system, transforms every infinitesimal circle C of the z-plane into an infinitesimal ellipse $\mathscr{E}$ the major axis of which is parallel to the u-axis, and the ratio of the semi-axes of which is a given function of the velocity of the flow λ, i.e., $p = p(\lambda)$; we note still that the flow velocity is equal to the ratio of the major semi-axis of $\mathscr{E}$ to the radius of the circle C.

As a third example, we will consider the motion of an ideal incompressible fluid in the presence of axial symmetry. It is known that in this case, if one selects as x-axis the axis of symmetry and as y-axis the distance from this axis, the velocity potential $u = u(x, y)$ of the flow satisfies the equation (cf., for example, M. A. Lavrent'ev, and B. V. Shabat, 1958, p. 298):

$$\frac{\partial^2 u}{\partial x^2} + \frac{1}{y}\frac{\partial u}{\partial y} + \frac{1}{y^2}\frac{\partial^2 u}{\partial y^2} = 0$$

[1] Cf., for example, M. A. Lavrent'ev and B. V. Shabat, Moscow, 1958, p. 311.

which is equivalent to the system

$$\frac{\partial u}{\partial x} = \frac{1}{y}\,\frac{\partial v}{\partial y},$$
$$\frac{\partial u}{\partial y} = -\,\frac{1}{y}\,\frac{\partial v}{\partial x}, \tag{4.9}$$

where v is the stream function.

As in the preceding example, the problem of finding an axi-symmetric flow around an axi-symmetric body is equivalent to the problem of the determination of a quasi-conformal mapping, corresponding to the system (4.9), of the axial cross-section of the flow region of the fluid on to a region bounded by straight lines parallel to the u-axis. The quasi-conformal mapping corresponding to the system (4.9) transforms any infinitesimal circle of the z-plane into an infinitesimal ellipse $\mathscr{E}$: the major axis of $\mathscr{E}$ is parallel to the u-axis for $y < 1$, parallel to the v-axis for $y > 1$, and the ratio of the semi-axes p is equal to $1/y$ for $y < 1$ and equal to y for $y > 1$.

For a geometric study of the general non-linear system (4.1), it is convenient to rewrite this system in a different form which we will call the e q u a t i o n s i n t e r m s o f c h a r a c t e r i s t i c s. For the purpose of the transition to this form, consider a continuously differentiable, homomorphic solution $f(z) = u(x, y) + iv(x, y)$ of the system (4.1) with non-zero Jacobian; the functions F_1 and F_2 will be assumed to be continuous together with their partial derivatives up to the third order inclusively. Fix a point x_0, y_0 and isolate the principal linear part (4.5) of the transformation under consideration at this point. Consider in the w-plane a unit square with vertex at the point $w_0 = u_0 + iv_0$ and with sides $\overline{w_0w_1}$, $\overline{w_0w_2}$, $w_2 - w_0 = (w_1 - w_2)e^{i\pi/2}$. Denote by ν the angle between the vector $\overline{w_0w_1}$ and the u-axis: $w_1 - w_0 = e^{i\nu}$. For the transformation (4.5), the unit square under consideration will correspond to some parallelogram Π_ν; let the points z_1, z_2 correspond in this process to the points w_1, w_2.

We write now

$$z_1 - z_0 = V_\nu e^{i\alpha_\nu}, \ \theta_\nu = \arg \frac{z_2 - z_0}{z_1 - z_0}, \tag{4.10}$$

$$W_\nu V_\nu J = 1,$$

where $J = \partial(u, v)/\partial(x, y)$ is the value of the functional determinant of the transformation (4.2) at the point z_0. The quantities V_ν, α_ν, θ_ν, W_ν (for any fixed value of ν) are completely determined by the parallelogram Π (fig. 18); we will call them the characteristics of the transformation $w = f(z)$ at the point $z = z_0$. These characteristics can be expressed in an elementary manner in terms of the partial derivatives of the functions u and v. In fact, introducing the usual notation

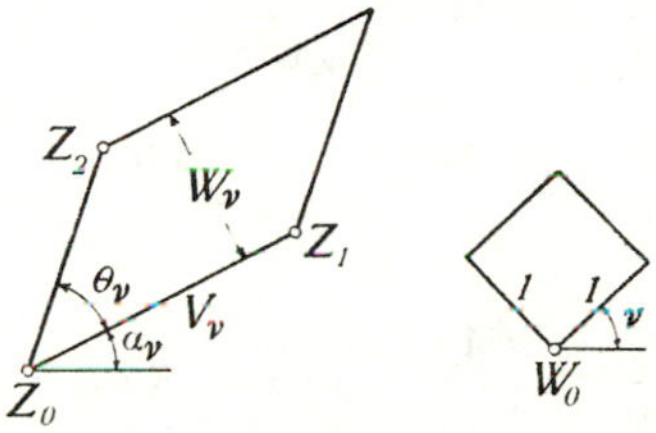

Fig. 18

$$E = \left(\frac{\partial u}{\partial x}\right)^2 + \left(\frac{\partial u}{\partial y}\right)^2, \quad F = \frac{\partial u}{\partial x}\frac{\partial v}{\partial x} + \frac{\partial u}{\partial y}\frac{\partial v}{\partial y},$$

$$G = \left(\frac{\partial v}{\partial x}\right)^2 + \left(\frac{\partial v}{\partial y}\right)^2,$$

we find

$$V_\nu = \frac{1}{J}\sqrt{E \sin^2 \nu - 2F \sin \nu \cos \nu + G \cos^2 \nu},$$

$$W_\nu = \frac{1}{\sqrt{E \sin^2 \nu - 2F \sin \nu \cos \nu + G \cos^2 \nu}},$$

$$\tan \alpha_\nu = -\frac{\dfrac{\partial v}{\partial x}\cos \nu - \dfrac{\partial u}{\partial x}\sin \nu}{\dfrac{\partial v}{\partial y}\cos \nu - \dfrac{\partial u}{\partial y}\sin \nu}, \tag{4.11}$$

$$\tan \theta_\nu = \frac{2J}{(E - G)\sin 2\nu - 2F \cos 2\nu}.$$

The formulae (4.11) can be solved for the partial derivatives in order to express these derivatives in terms of these characteristics:

$$\begin{aligned}
\frac{\partial u}{\partial x} &= \left(\frac{\tan \alpha_\nu + \tan \theta_\nu}{V_\nu \tan \theta_\nu} + \frac{\tan \alpha_\nu \tan \nu}{W_\nu}\right) \cos \alpha_\nu \cos \nu, \\
\frac{\partial u}{\partial y} &= \left(\frac{\tan \alpha_\nu \tan \theta_\nu - 1}{V_\nu \tan \theta_\nu} - \frac{\tan \nu}{W_\nu}\right) \cos \alpha_\nu \cos \nu, \\
\frac{\partial v}{\partial x} &= \left(\frac{\tan \alpha_\nu + \tan \theta_\nu}{V_\nu \tan \theta_\nu} \tan \nu - \frac{\tan \alpha_\nu}{W_\nu}\right) \cos \alpha_\nu \cos \nu, \\
\frac{\partial v}{\partial y} &= \left(\frac{\tan \alpha_\nu \tan \theta_\nu - 1}{V_\nu \tan \theta_\nu} \tan \nu + \frac{1}{W_\nu}\right) \cos \alpha_\nu \cos \nu.
\end{aligned} \tag{4.12}$$

Substituting the last expressions into (4.1), we find the relations

$$\begin{aligned}
G_1(x, y, u, v, V_\nu, W_\nu, \alpha_\nu, \theta_\nu) &= 0, \\
G_2(x, y, u, v, V_\nu, W_\nu, \alpha_\nu, \theta_\nu) &= 0
\end{aligned} \tag{4.13}$$

which will be called the equations in terms of characteristics. We will assume that the equations in characteristics (for any ν) can be solved for W_ν and θ_ν

$$\begin{aligned}
W_\nu &= W_\nu(x, y, u, v, V_\nu, \alpha_\nu), \\
\theta_\nu &= \theta_\nu\ (x, y, u, v, V_\nu, \alpha_\nu).
\end{aligned} \tag{4.14}$$

In certain problems, the characteristics need only be considered for $\nu = 0$; omitting in this case the subscripts, we will denote the characteristics by V, W, α and θ, respectively.

4.2. Derivative systems

For the proof of the existence theorem as well as for the study of the properties of quasi-conformal transformations, it is very important to extend to such mappings the concept of the derivative theory of analytic functions. If a system coincides with the Cauchy-Riemann equations, the characteristics $P = \log 1/V = \log |f'(z)|$ and $\alpha = \arg f'(z)$ are conjugate harmonic functions of the variables x, y as well as of the variables u, v:

$$\frac{\partial P}{\partial v} = -\frac{\partial \alpha}{\partial u}, \quad \frac{\partial \alpha}{\partial v} = \frac{\partial P}{\partial u}.$$

We will show that in the general case the characteristics $P = \log 1/V$ and α (for $\nu = 0$) satisfy the system of equations

$$\begin{aligned} \frac{\partial P}{\partial v} &= a_1 \frac{\partial P}{\partial u} + a_2 \frac{\partial \alpha}{\partial u} + a_3, \\ \frac{\partial \alpha}{\partial v} &= b_1 \frac{\partial P}{\partial u} + b_2 \frac{\partial \alpha}{\partial u} + b_3, \end{aligned} \tag{4.15}$$

the coefficients of which are given by the formulae

$$\begin{aligned} a_1 &= \frac{\partial W}{\partial V} \cot \theta - \frac{\partial \theta}{\partial V} \frac{W}{\sin^2 \theta}, \\ a_2 &= \frac{1}{V}\left(-\frac{\partial W}{\partial \alpha} \cot \theta + \frac{\partial \theta}{\partial \alpha} \frac{W}{\sin^2 \theta} + W\right), \\ b_1 &= -\frac{\partial W}{\partial V}, \quad b_2 = \frac{1}{V}\left(\frac{\partial W}{\partial \alpha} + W \cot \theta\right), \\ a_3 &= -\left(\frac{1}{V} \frac{\partial W}{\partial u} + \frac{\partial W}{\partial s}\right) \cot \theta + \left(\frac{1}{V} \frac{\partial \theta}{\partial u} + \frac{\partial \theta}{\partial s}\right) \frac{W}{\sin^2 \theta}, \\ b_3 &= \frac{1}{V} \frac{\partial W}{\partial u} + \frac{\partial W}{\partial s}, \end{aligned} \tag{4.16}$$

where $\partial/\partial s = \partial/\partial x \cos \alpha + \partial/\partial y \sin \alpha$ is the derivative in the direction of the base of the parallelogram Π. We will call the system (4.15) the d e r i v a t i v e s y s t e m of the system (4.1).

Consider a geometric derivation of the derivative system. We introduce the following assumptions:

1°. the right-hand sides of the equations in characteristics (for $\nu = 0$)

$$W = W(x, y, u, v, V, \alpha), \quad \theta = \theta(x, y, u, v, V, \alpha) \tag{4.17}$$

have continuous first order partial derivatives in some region of the six-dimensional space;

2°. the solution $w = f(z)$ under consideration accomplishes a homomorphic transformation of the region D on to Δ; it has everywhere in D continuous second order partial derivatives and a positive Jacobian $J = \partial(u, v)/\partial(x, y)$;

3°. all the values of x, y, u, v, V, α belong for the solution under consideration to the region referred to in 1°.

Consider in the region Δ an infinitesimal square q with sides parallel to the coordinate axes; let the length of the sides of the square be equal to du. In the case of the transformation under consideration, the square q corresponds to some curvilinear quadrangle Q in the region D. Let AB and CD be the sides of Q corresponding to the sides ab and cd of the square q, parallel to the u-axis.

In order to derive the first equation of the system (4.15), it is sufficient to compute the principal second order part of the difference of the lengths of the sides CD and AB. Since the functions u and v have, by assumption, continuous second order derivatives, the curvatures of the sides AB and CD, and likewise of the sides AC and BD will be infinitely close to each other. Hence it is readily deduced that for the purpose of our computations the curvilinear quadrangle can be replaced by a rectilinear quadrangle Q_1: $AB_1C_1D_1$ (fig. 19) with the sides

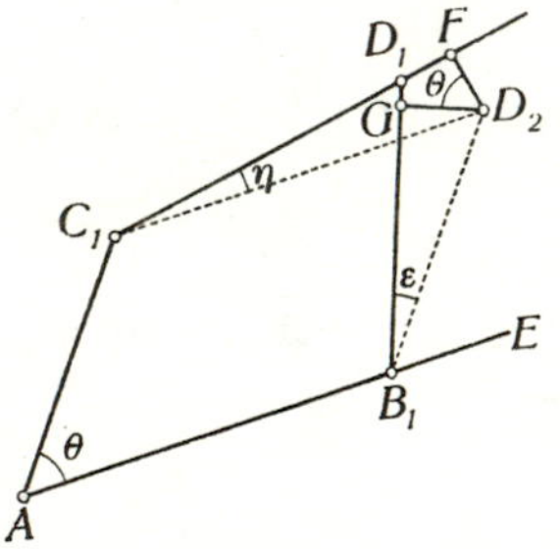

Fig. 19

$$AB_1 = V\,du, \quad \sphericalangle C_1AB_1 = \theta,$$

$$AC_1 = \frac{W\,du}{\sin\theta}, \quad B_1D_1 = \frac{W + dW}{\sin(\theta + d\theta)},$$

$$\sphericalangle D_1B_1E = \theta + d\theta + d\alpha,$$

where dW, $d\theta$, $d\alpha$ are the total increments of the functions W, θ, α for constant v, when u undergoes an increase du. Since we have for $v = $ const. that $\partial x/\partial u = V \cos\alpha$, $\partial y/\partial u = V \sin\alpha$, we find

$$dW = \left(\frac{\partial W}{\partial x} V \cos \alpha + \frac{\partial W}{\partial y} V \sin \alpha + \right.$$

$$\left. + \frac{\partial W}{\partial u} + \frac{\partial W}{\partial V} \frac{\partial V}{\partial u} + \frac{\partial W}{\partial \alpha} \frac{\partial \alpha}{\partial u}\right) du,$$

$$d\theta = \left(\frac{\partial \theta}{\partial x} V \cos \alpha + \frac{\partial \theta}{\partial y} V \sin \alpha + \right.$$

$$\left. + \frac{\partial \theta}{\partial u} + \frac{\partial \theta}{\partial V} \frac{\partial V}{\partial u} + \frac{\partial \theta}{\partial \alpha} \frac{\partial \alpha}{\partial u}\right) du,$$

$$d\alpha = \frac{\partial \alpha}{\partial u} du.$$

For the determination of the difference $C_1D_1 - AB_1$ we construct the parallelogram with the sides AB_1 and AC_1; let D_2 be the vertex of this parallelogram which is opposite to the vertex A. Draw from D_2 the perpendiculars D_2F on to C_1D_1 and D_2G on to B_1D_1. Neglecting infinitesimal higher order quantities, we find

$$C_1D_1 - AB_1 = \frac{\partial V}{\partial v} dv \cdot du = \frac{\partial V}{\partial v} du^2 = - D_1F =$$

$$= - (D_2G \sin \theta - D_1G \cos \theta)$$

(within the degree of accuracy adopted, we can assume that $D_2F \perp C_1D_2$ and $D_2G \perp B_1D_2$). However, since $\sin \varepsilon = \varepsilon = = d\theta + d\alpha$ (fig. 19), we obtain

$$D_2G = B_1D_2 \sin \varepsilon = \frac{W du}{\sin \theta} (d\theta + d\alpha)$$

and

$$D_1G = B_1D_1 - B_1D_2 = \left(\frac{W + dW}{\sin (\theta + d\theta)} - \frac{W}{\sin \theta}\right) du =$$

$$= \frac{\sin \theta \, dW - W \cos \theta d\theta}{\sin^2 \theta} du.$$

Substituting these results into the expression for $C_1D_2 - AB_1$, we find

$$\frac{\partial V}{\partial v}\,du^2 = [\cotan\theta\,dW - W\cotan^2\theta\,d\theta - W(d\theta + d\alpha)]\,du =$$

$$= \left(\cotan\theta\,dW - \frac{W}{\sin^2\theta}\,d\theta - W\,d\alpha\right)du.$$

Finally, after substitution of the above expressions for dW, $d\theta$, $d\alpha$, we arrive at the first equation of the system (4.15).

For the purpose of the derivation of the second equation of the system, we note that

$$\frac{\partial\alpha}{\partial v}\,dv = \frac{\partial\alpha}{\partial v}\,du = \eta = \sin\eta = \frac{D_2F}{C_1D_2} = \frac{D_2F}{AB_1},$$

(cf. fig. 19). However,

$$AB_1 = V\,du,$$

$$D_2F = D_1G\sin\theta + D_2G\cos\theta = (dW + W\cotan\theta\,d\alpha)\,du$$

(where we have used the above expressions for D_1G and D_2G), and after obvious manipulations we find the second equation of the system (4.15).

Note the particular case when the derivative system assumes its simplest form. If the equations in terms of the characteristics (4.17) do not contain explicitly the coordinates x, y, u, v, then, as can be seen from (4.16), one has $a_3 = b_3 = 0$ and the derivative system will be homogeneous and linear with respect to the partial derivatives

$$\begin{aligned}\frac{\partial P}{\partial v} &= a_1\frac{\partial P}{\partial u} + a_2\frac{\partial\alpha}{\partial u},\\ \frac{\partial\alpha}{\partial v} &= b_1\frac{\partial P}{\partial u} + b_2\frac{\partial\alpha}{\partial u},\end{aligned}\qquad(4.18)$$

where the coefficients a_1, a_2, b_1, b_2 depend only on P and α. If we select in this system P and α as independent variables and assume u and v to be unknown functions, our system becomes after some obvious transformations a homogeneous, linear system.

This circumstance applies, in particuliar, for the equations of gas dynamics. In fact, the equations of gas dynamics (in case of plane steady motion of an ideal gas) written in terms of the characteristics have the form

$$W = W(V), \quad \theta \equiv \frac{\pi}{2} \tag{4.19}$$

(the first equation expresses the gas regime: the relationship between the transport $1/W$ and the flow velocity $1/V$). In this case, the derivative becomes

$$\begin{aligned} \frac{\partial P}{\partial v} &= \frac{W}{V} \frac{\partial \alpha}{\partial u}, \\ \frac{\partial \alpha}{\partial v} &= -\frac{dW}{dV} \frac{\partial P}{\partial u}. \end{aligned} \tag{4.20}$$

The first to obtain this system for the equations of gas dynamics was S. A. Chaplygin and it is called after him. Chaplygin's system forms the basis of the hodograph method which has found important applications in gas dynamics.

4.3. Strong ellipticity

With the objective of extending to quasi-conformal transformations as large a number of the geometric theorems of the theory of conformal mappings as possible, we limit ourselves here to the study of classes of quasi-conformal transformations which correspond to elliptic systems. In the simplest case of a system which is linear and homogeneous with respect to the derivatives

$$\frac{\partial u}{\partial x} = a_{11} \frac{\partial v}{\partial x} + a_{12} \frac{\partial v}{\partial y}, \quad \frac{\partial u}{\partial y} = a_{21} \frac{\partial v}{\partial x} + a_{22} \frac{\partial v}{\partial y}, \tag{4.21}$$

where the coefficients a_{ik} are given functions of x, y, u, v and the condition of ellipticity

$$\left\{\frac{a_{11} - a_{22}}{2}\right\}^2 + a_{12}a_{21} < 0 \tag{4.22}$$

ensures topological equivalence of the solutions to conformal transformations, preservation of Riemann's theorem on the existence and uniqueness of mappings and a number of other important properties. These results are contained in studies by the author, Z. Ia. Shapiro, B. V. Shabat, C. B. Morrey, I. N. Vekua, G. N. Polozhii, L. Bers, L. Bers and L. Nirenberg and B. V. Boiarskii.

However, very simple examples show that even in the case of linear non-homogeneous systems, and more so in the case of non-linear systems, the usual analytic condition of ellipticity does not ensure similar results (cf. B. V. Shabat, 1957). Therefore, we will derive a geometric condition which we will call the condition of strong ellipticity and which makes it possible to extend to (generally speaking, non-linear) systems satisfying this condition a number of important theorems of the theory of conformal transformations.

We will say that the system (4.1) is s t r o n g l y e l l i p t i c in some region D of the space $(x, y, u, v, \partial u/\partial x, \partial u/\partial y, \partial v/\partial x, \partial v/\partial y)$ if for any ν $(0 \leq \nu < 2\pi)$ and for any values of the arguments in D this system can be written in the form of the equations in terms of the characteristics (4.14), where

1°. the functions

$$W_\nu(x, y, u, v, V_\nu, \alpha_\nu) \text{ and } \theta_\nu(x, y, u, v, V_\nu, \alpha_\nu)$$

are single-valued and differentiable for all values of the arguments corresponding to the points of D;

2°. there exists a constant $k > 0$ such that for all such values of the arguments

$$k < \theta_\nu(x, y, u, v, V_\nu, \alpha_\nu) < \pi - k; \tag{4.23}$$

3°. there exists a constant $k_1 > 0$ such that for those values of the arguments

$$\frac{\partial W_\nu}{\partial V_\nu} > k_1. \tag{4.24}$$

For $k = k_1 = 0$, the system will be said to be simply elliptic (in the geometric sense).

We will now explain the link between this condition and the ordinary condition of ellipticity. Consider first the case of a system of the form (4.21) which is linear and homogeneous with respect to the partial derivatives. Then we will find for such systems equations in terms of characteristics for $\nu = 0$. For $\nu = 0$, we obtain from (4.11)

$$V = \frac{\sqrt{G}}{J}, \; W = \frac{1}{\sqrt{G}}, \; \tan\alpha = -\frac{\dfrac{\partial v}{\partial x}}{\dfrac{\partial v}{\partial y}}, \; \tan\theta = -\frac{J}{F}, \tag{4.25}$$

where the last formula can be rewritten in the form

$$\sin\theta = \frac{J}{\sqrt{F^2 + J}} = \frac{J}{\sqrt{EG}}.$$

Replacing the derivatives of the function u by derivatives of the function v in accordance with the system (4.21), and setting then $\partial v/\partial x = -(\partial v/\partial y)\tan\alpha$, we find the expression for the Jacobian

$$J = -[a_{21}\tan^2\alpha + (a_{11} - a_{22})\tan\alpha - a_{12}]\left(\frac{\partial v}{\partial y}\right)^2.$$

For $\nu = 0$, we derive from (4.12) that $\partial v/\partial y = (\cos\alpha)/W$; substituting this result into the expression for J and letting $J = 1/VW$, we find the first equation in characteristics

$$W = [a_{12}\cos^2\alpha + (a_{22} - a_{11})\sin\alpha\cos\alpha - a_{21}\sin^2\alpha]V. \tag{4.26}$$

In a similar manner, we obtain the second equation

$$\sin\theta = \frac{J}{\sqrt{EG}} =$$

$$= \frac{a_{12}\cos^2\alpha + (a_{22} - a_{11})\sin\alpha\cos\alpha - a_{21}\sin^2\alpha}{\sqrt{(a_{11}\sin\alpha - a_{12}\cos\alpha)^2 - (a_{21}\sin\alpha - a_{22}\cos\alpha)^2}}. \tag{4.27}$$

It is seen from these equations that for $\nu = 0$ the conditions of ellipticity (4.23) and (4.24) are equivalent in a geometric sense to the condition of positive definiteness of the quadratic form in (4.26), i.e., they reduce to the inequality

$$\left\{\frac{a_{11} - a_{22}}{2}\right\}^2 + a_{12}a_{21} < 0.$$

By virtue of the invariance of this inequality with respect to rotations, it is equivalent also to the geometric condition of ellipticity for any ν. The condition of strong ellipticity is equivalent to the condition that the minimum of the quadratic form in (4.26) should be larger than some positive constant.

In the non-linear (or non-homogeneous) case, the geometric condition of ellipticity does not reduce to the ordinary definition. B. V. Shabat has shown that in the case when the equations in terms of characteristics do not involve the coordinates explicitly, the condition reduces to the ordinary condition of ellipticity, supplemented by the condition that the Jacobian of any solution of the system must vanish only when all its partial derivatives vanish, and that no change of sign must occur.

It is not difficult to verify by simply examples that in the general case of non-linear systems the condition $\partial W/\partial V > 0$ does not entail the condition $\partial W_\nu/\partial V_\nu > 0$ for any ν.

In conclusion of this chapter, we will establish a basic theorem linking the concept of geometric ellipticity to that of the derivative system.

THEOREM 4.1. *For the non-linear system* (4.1) *let the equations in characteristics*

$$W = W(x, y, u, v, V, \alpha), \quad \theta = \theta(x, y, u, v, V, \alpha) \qquad (4.17)$$

in some region of the six-dimensional space have continuous first order partial derivatives. Assume also that (4.1) *does not belong to the exclusive case of systems for which*

$$V_\nu^2 \frac{\partial W_\nu}{\partial V_\nu} = V^2 \frac{\partial W}{\partial V}$$

for all ν. *Then the condition that the derivative* $\partial W_\nu / \partial V_\nu$ *is to be positive for any* ν *is equivalent to the condition of ellipticity of the derivative system* (4.15) [1].

In order to prove this theorem, we find first of all from (4.12) for $\nu = 0$

$$E = \frac{1}{V^2 \sin^2 \theta}, \quad F = -\frac{1}{WV \tan \theta}, \quad G = \frac{1}{W^2}$$

and then, by (4.11),

$$W_\nu = \frac{WV \sin \theta}{\sqrt{W^2 s^2 + 2scWV \sin \theta \cos \theta + V^2 c^2 \sin^2 \theta}},$$

$$V_\nu = \frac{1}{\sin \theta} \sqrt{W^2 s^2 + 2scWV \sin \theta \cos \theta + V^2 c^2 \sin^2 \theta}, \tag{4.28}$$

$$\tan \alpha_\nu = \frac{V \tan \alpha + W\left(1 + \dfrac{\tan \alpha}{\tan \theta}\right) \tan \nu}{V - W\left(\tan \alpha - \dfrac{1}{\tan \theta}\right) \tan \nu},$$

where $c = \cos \nu$, $s = \sin \nu$. We will assume that in these formulae W and θ have been replaced by their expressions in terms of characteristics; solving the second and third of these equations with respect to V and α, we obtain

$$V = g_1(x, y, u, v, V_\nu, \alpha_\nu), \quad \alpha = g_2(x, y, u, v, V_\nu, \alpha_\nu)$$

and, after substitution of these expressions into the first equation, the equation in characteristics for any ν:

$$W_\nu = W_\nu(x, y, u, v, V_\nu, \alpha_\nu).$$

The condition that this system can be solved in the manner

---

[1] The proof of this theorem was given by B. V. Shabat in 1961

indicated is that the following determinant must be non-zero:[1]

$$D = \frac{\partial(V_\nu, \alpha_\nu)}{\partial(V, \alpha)} = \frac{V}{V_\nu}\Big\{\cos^2\nu +$$

$$+\left[\frac{1}{V}\left(W\operatorname{cotan}\theta + \frac{\partial W}{\partial\alpha}\right) + \frac{\partial W}{\partial V}\operatorname{cotan}\theta - \frac{W}{\sin^2\theta}\frac{\partial\theta}{\partial V}\right]\cos\nu\sin\nu +$$

$$+\frac{W}{V\sin^2\theta}\left[\frac{\partial(W,\theta)}{\partial(V,\alpha)} + \frac{\partial W}{\partial V} - W\operatorname{cotan}\theta\frac{\partial\theta}{\partial V}\right]\sin^2\nu\Big\};$$

substituting here the expressions for the coefficients of the derivative system (4.16), we will have

$$D = \frac{V}{V_\nu}[\cos^2\nu + (a_1+b_2)\cos\nu\sin\nu + (a_1b_2 - a_2b_1)\sin^2\nu]. \tag{4.29}$$

By the rule for differentiation of implicit functions, we obtain

$$\frac{\partial g_1}{\partial V_\nu} = \frac{1}{D}\left(\frac{\partial\alpha_\nu}{\partial W}\frac{\partial W}{\partial\alpha} + \frac{\partial\alpha_\nu}{\partial\theta}\frac{\partial\theta}{\partial\alpha} + \frac{\partial\alpha_\nu}{\partial\alpha}\right),$$

$$\frac{\partial g_2}{\partial V_\nu} = -\frac{1}{D}\left(\frac{\partial\alpha_\nu}{\partial W}\frac{\partial W}{\partial V} + \frac{\partial\alpha_\nu}{\partial V} + \frac{\partial\alpha_\nu}{\partial\theta}\frac{\partial\theta}{\partial V}\right),$$

and hence

$$\frac{\partial W_\nu}{\partial V} = \left(\frac{\partial W_\nu}{\partial W}\frac{\partial W}{\partial V} + \frac{\partial W_\nu}{\partial V} + \frac{\partial W_\nu}{\partial\theta}\frac{\partial\theta}{\partial V}\right)\frac{\partial g_1}{\partial V_\nu} +$$

$$+\left(\frac{\partial W_\nu}{\partial V}\frac{\partial W}{\partial\alpha} + \frac{\partial W_\nu}{\partial\theta}\frac{\partial\theta}{\partial\alpha}\right)\frac{\partial g_2}{\partial V_\nu}.$$

Substituting here the expressions for $\partial W_\nu/\partial W, \cdots$ found from (4.28) by use of (4.17), we obtain, finally,

$$\frac{\partial W_\nu}{\partial V_\nu} = \frac{1}{V_\nu^3 D}\Big\{V^3\frac{\partial W}{\partial V}\cos^2\nu + WV\left(\frac{\partial W}{\partial\alpha} + W\operatorname{cotan}\theta +\right.$$

---

[1] In order to obtain this expression, we compute the derivatives $\partial V_\nu/\partial V, \cdots$ from (4.28) after substitution of (4.27).

$$+ \frac{WV}{\sin^2\theta}\,\frac{\partial\theta}{\partial V} + V\frac{\partial W}{\partial V}\operatorname{cotan}\theta\Big)\cos\nu\sin\nu +$$

$$+ \frac{W^3}{\sin^2\theta}\left(1 + \frac{\partial\theta}{\partial\alpha} + V\frac{\partial\theta}{\partial V}\operatorname{cotan}\theta\right)\sin^2\nu\Big\}. \qquad (4.30)$$

The theorem is now readily proved. Let the derivative system (4.15) be elliptic, i.e., for all admissible values of the arguments, let there be fulfilled the inequality

$$4a_2b_1 + (a_1 - b_2)^2 < 0.$$

However, this is the condition for the quadratic form in (4.29) to be definite; obviously, this form assumes positive values, and hence $D \geq 0$. Further, elementary manipulations using the formulae (4.16) for the coefficients of the derivative system show that the discriminant of the quadratic form in the braces in (4.30) is equal to $V^4W^2[4a_2b_1 + (a_1 - b_2)^2]$, i.e., that it is positive. Consequently, the expression in braces is positive, and therefore also $\partial W_\nu/\partial V_\nu$ is positive for all ν, if $\partial W/\partial V > 0$.

Conversely, let $\partial W_\nu/\partial V_\nu > 0$ for all ν, then, as is seen from (4.30), D and the form in braces have the same value for all ν. Thus, either these forms are definite or their ratio does not depend on ν. For the first case, it has been shown by the preceding manipulations that the condition of definiteness of these quadratic forms coincides with the condition of ellipticity of the derivative system. The second case is shown to be exclusive, by (4.30), since it reduces to systems for which for all ν

$$V_\nu^2\,\frac{\partial W_\nu}{\partial V_\nu} = V^2\frac{\partial W}{\partial V}.$$

CHAPTER V

LINEAR SYSTEMS

5.1. Transformations with bounded distortion

Let there be given in some region D the homomorphic differentiable transformation with a non-zero Jacobian

$$w = f(z) = u(x, y) + iv(x, y). \tag{5.1}$$

At every point z of this region, the principal linear part of the transformation $f(z)$

$$U - u = u_x(X - x) + u_y(Y - y),$$
$$V - v = v_x(X - x) + v_y(Y - y)$$

maps the homothetic ellipse

$$(u_x^2 + v_x^2)(X - x)^2 + (u_x u_y + v_x v_y)(X - x)(Y - y) + \\ + (u_y^2 + v_y^2)(Y - y)^2 = r^2$$

into the circle $(U - u)^2 + (V - v)^2 = r^2$. If the ratio of semi-axes $p = p(z) \geq 1$ of these ellipses is bounded in D from above by some constant, the transformation $f(z)$ is said to have bounded distortion. Elementary computations show that by the condition of bounded distortion there exists a constant $Q = q + \frac{1}{2} \geq 2$ such that everywhere in D

$$E + G \leq QJ, \tag{5.2}$$

where

$$E = u_x^2 + u_y^2, \; G = v_x^2 + v_y^2, \; J = u_x v_y - u_y v_x.$$

If (5.1) is a quasi-conformal transformation corresponding to a strongly elliptic system, we have, by (4.12),

$$\frac{E+G}{J} = \frac{1}{\sin\theta}\left(\frac{W}{V\sin\theta} + \frac{V\sin\theta}{W}\right).$$

Therefore, for the entire class of quasi-conformal mappings corresponding to strongly elliptic systems, the distortion is bounded by a constant which depends only on the system under consideration.

Thus, the study of transformations with bounded distortion is of special interest for the general theory. The properties of such mappings have great independent significance; they also find wide application in the geometric theory of functions of a complex variable.

We will derive here certain basic properties of mappings with bounded distortion. Since the property of boundedness of distortion is obviously invariant with respect to conformal transformations, we can consider for the study of mappings with bounded distortion in simply connected regions, without restricting generality, mappings of the circle $|z| < 1$ on to the circle $|w| < 1$ with the normalization condition $f(0) = 0$.

5.1.1. Equi-graded continuity. We will begin with two auxiliary propositions.

LEMMA 5.1. *For the quasi-conformal transformation $w = f(z)$, $f(0) = 0$, of the circle $|z| < 1$ on to the circle $|w| < 1$ with bounded distortion $p(z) \leq q$, let the circle $|z| < r$ go over into the region Δ_r with area $S = S(r)$; then we have for any r, $0 \leq r \leq 1$, that*

$$S(r) \leq \pi r^{2/q}. \tag{5.3}$$

For the proof, we denote by $L(r)$ the length of the boundary L_r of the region Δ_r and obtain an estimate of dS; we have

$$dS \geq dr \int_0^{2\pi} \left(\frac{dL}{ds}\right)^2 \frac{r}{p}\, d\varphi,$$

where dL and $ds = r d\varphi$ are the corresponding elements of length of the arc L_r and the circle $|z| = r$. Hence, replacing p by $q \geq p$ and using the Buniakovskii-Schwarz inequality, we obtain

$$dS \geq \frac{dr}{q} \frac{1}{2\pi r} \left(\int_0^{2\pi} \frac{dL}{d\varphi} d\varphi \right)^2 = \frac{dr}{q} \frac{L^2(r)}{2\pi r};$$

however, by a well-known isoperimetric property of the circle, one has

$$L^2 \geq 4\pi S,$$

and, consequently,

$$dS \geq \frac{2}{q} S \frac{dr}{r}.$$

Therefore, letting $S = S_0$ for $r = r_0$, we find

$$S \geq S_0 \left(\frac{r}{r_0} \right)^{2/q}.$$

But we have $S = \pi$ for $r = 1$; consequently, $\pi \geq S_0(1/r_0)^{2/q}$, whence, replacing the arbitrary r_0 by r, we derive the required inequality (5.3), and the lemma has been proved.

LEMMA 5.2. *Under the conditions of Lemma* 5.1, *we have*

$$|f(z)| < k|z|^{1/q}, \tag{5.4}$$

where k is a constant which does not depend on z.

In fact, retaining the notation above, we denote the diameter of the region Δ_r by h_r. We will now map the region bounded by L_r and the circle $|w| = 1$ conformally on to the ring $\rho < |\zeta| < 1$:

$$\zeta = \varphi(w).$$

In accordance with the properties of conformal transformations, we have

$$\rho \geq \frac{1}{k} h_r,$$

where k is some constant.

Consider now the implicit function

$$\zeta = \varphi[f(z)];$$

it gives a conformal mapping of the ring $r < |z| < 1$ on to the ring $\rho < |\zeta| < 1$, where the characteristic $p(z)$ of this transformation coincides with that of the basic transformation; in particular, one has $p(z) \leq q$. However, by Lemma 5.1,

$$\pi\rho^2 \leq \pi r^{2/q};$$

hence

$$|f(z)| \leq h_r \leq kr^{1/q},$$

and Lemma 5.2 has been proved.

From these Lemmas follows the first basic theorem relating to mappings with bounded distortion, the theorem on equigraded continuity:

THEOREM 5.1. *If the function*

$$w = f(z), \quad f(0) = 0$$

quasi-conformally maps the circle $|z| < 1$ *on to the circle* $|w| < 1$ *with bounded distortion* $p(z) \leq q$, *we have for any pair of points* z_1 *and* z_2

$$|f(z_2) - f(z_1)| \leq K|z_2 - z_1|^{1/q}, \tag{5.5}$$

where K is a constant which does not depend on z.

For the purpose of the proof, we spread our mapping, first of all, over the circles $|z| < 2$ and $|w| < 2$, where the characteristic p is still to remain smaller than q. However, in that case, the mapping

$$w = \tfrac{1}{2}f(2z) = F(z)$$

will satisfy the conditions of the theorem, where it is sufficient for the purpose of the unknown estimate (5.5) to obtain an analogous estimate for $F(z)$ for $|z_1| < \frac{1}{2}$, $|z_2| < \frac{1}{2}$.

We now map conformally the circle $|z| < 1$ on to the circle $|\zeta| < 1$:

$$z = \Phi(\zeta), \quad \Phi(0) = z_1,$$

and the circle $|w| < 1$ on to the circle $|\omega| < 1$:

$$\omega = \Psi(w), \quad \Psi[f(z_1)] = 0.$$

Then we can apply to the transformation

$$\omega = \Psi\{f[\Phi(\zeta)]\}$$

Lemma 5.2 which renders

$$|\Delta\omega| \le k|\Delta\zeta|^{1/q};$$

however, since for the fractionally linear transformations Φ and Ψ the moduli of the derivatives are bounded from above and from below by positive constants, the ratios $|\Delta w/\Delta\omega|$ and $|\Delta\zeta/\Delta z|$ will also be bounded, and this result leads to the required estimate (5.5).

5.1.2. Almost conformal mappings. A second essential result of the theory of mappings with bounded distortion is the stability of these transformations: small changes of the characteristics also cause small changes of the transformation. We will present an exact formulation of this important property in the next section; at this stage we will study the case of a transformation which is almost conformal

THEOREM 5.2. *Let the transformation*

$$w = f(z), \ f(0) = 0, \ f(1) = 1$$

of the circle $|z| < 1$ on to the circle $|w| < 1$ be almost conformal in the sense that everywhere on the circle

$$1 \le p(z) \le 1 + \varepsilon; \tag{5.6}$$

then

$$|f(z) - z| < \lambda(\varepsilon), \tag{5.7}$$

where $\lambda(\varepsilon)$ is a universal function, $\lim_{\varepsilon\to 0} \lambda(\varepsilon) = 0$. [1]

---

[1] P. P. Belinskii (1956), using a more intricate reasoning, has shown that one can take

$$\lambda(\varepsilon) = 18\varepsilon.$$

For the proof we use the formal derivatives

$$\frac{\partial}{\partial \bar{z}} = \frac{1}{2}\left(\frac{\partial}{\partial x} + i\,\frac{\partial}{\partial y}\right), \quad \frac{\partial}{\partial z} = \frac{1}{2}\left(\frac{\partial}{\partial x} - i\,\frac{\partial}{\partial y}\right)$$

(cf., for example, M. A. Lavrent'ev and B. V. Shabat, 1958, p. 295); it is readily seen that we have in the case of the quasi-conformal mapping $w = f(z)$

$$\left|\frac{\partial f}{\partial \bar{z}}\right| = \frac{p-1}{p+1}\left|\frac{\partial f}{\partial z}\right|$$

(cf. also the next section), and hence the Jacobian of the transformation

$$J = \left|\frac{\partial f}{\partial z}\right|^2 - \left|\frac{\partial f}{\partial \bar{z}}\right|^2 = \frac{4p}{(p-1)^2}\left|\frac{\partial f}{\partial \bar{z}}\right|^2.$$

Let C be an arbitrary closed rectilinear contour belonging to the circle $|z| < 1$; by the Riemann-Green formula in terms of the formal derivatives, we have

$$\left|\int_C f(z)\,dz\right| = \left|2i \iint_{|z|<1} \frac{\partial f}{\partial \bar{z}}\,dx\,dy\right| \leq 2 \iint_{|z|<1} \frac{p-1}{2\sqrt{p}}\sqrt{J}\,dx\,dy.$$

Using the condition (5.6) of the theorem and the Buniakovskii-Schwarz formula, we find therefore

$$\left|\int_C f(z)\,dz\right| \leq \varepsilon \iint_{|z|<1} \sqrt{J}\,dx\,dy \leq \varepsilon \sqrt{\pi \iint_{|z|<1} J\,dx\,dy} = \varepsilon\pi. \tag{5.8}$$

By Theorem 5.1, one has for $p \leq 1 + \varepsilon$ a family of functions $\{f(z)\}$; in the same way, there is also a manifold of inverse functions with equal degree of continuity, where we will have for every function $f_0(z)$ in the limit $\varepsilon \to 0$, by (5.8),

$$\int_C f_0(z)\,dz = 0,$$

i.e., by Morera's theorem, $f_0(z)$ is an analytic function.

Since $f_0(z)$ renders a single-valued mapping of the unit circle on to itself and $f_0(0) = 0$, $f_0(1) = 1$, we have $f_0(z) \equiv z$. Hence, using still the equal degree of continuity of our manifold, we obtain that $f(z)$ tends uniformly to z for $\varepsilon \to 0$; this proves the theorem.

We will still give one more result which proves that closeness of a mapping to a conformal mapping in the small (i.e., proximity of the characteristic p to 1) implies also closeness as a whole.

THEOREM 5.3. *Let $w = f(z)$, $f(0) = 0$ be a quasi-conformal mapping of the circle $|z| < 1$ on to the circle $|w| < 1$, where $p(z) \leq 1 + \varepsilon$; under these conditions there exists a universal function $\eta(\varepsilon, \varepsilon_1)$*

$$\lim_{\substack{\varepsilon \to 0 \\ \varepsilon_1 \to 0}} \eta(\varepsilon, \varepsilon_1) = 0,$$

such that for any point z_0 of the circle $|z| < 1$, any ρ: $0<\rho<1-|z_0|$, and two arbitrary points z_1 and z_2 of the ring $(1-\varepsilon_1)\rho<|z-z_0|<\rho$, we have

$$\left| \left| \frac{f(z_2) - f(z_0)}{f(z_1) - f(z_0)} \right| - 1 \right| < \eta(\varepsilon, \varepsilon_1), \tag{5.9}$$

$$\left| \arg \frac{f(z_2) - f(z_0)}{f(z_1) - f(z_0)} - \arg \frac{z_2 - z_0}{z_1 - z_0} \right| < \eta(\varepsilon, \varepsilon_1). \tag{5.10}$$

We will restrict the proof to the first inequality, because the second can be derived in an analogous manner. Using the reasoning leading to the proof of Theorem 5.1, we can limit ourselves to the case of a ring with centre at the point 0, $z_0 = 0$. We will assume that the theorem is not true and arrive at a contradiction.

Denote by Δ_ρ the image of the ring $(1 - \varepsilon_1)\rho < |z| < \rho < 1$ in the w-plane; by assumption, there exists a number $k > 1$ such that, whatever may be the small numbers ε and ε_1, there exists a function $w = f(z)$, which satisfies Lemmas 1 and 2, and a region Δ_ρ, $\rho = \rho(\varepsilon, \varepsilon_1)$ containing points w_1 and w_2 for which

$$\left| \frac{w_2}{w_1} \right| > k. \tag{5.11}$$

By Theorem 5.2, it is sufficient to deal with the case

$$\lim_{\substack{\varepsilon\to 0\\ \varepsilon_1\to 0}} \rho = 0.$$

Construct in the z-plane a circle $|z| = n\rho = \rho_1 < 1$, where n is a positive, sufficiently large number. Denote by D_{ρ_1} the image of the circle $|z| < \rho_1$ in the w-plane and let r be the distance of the point $w = 0$ from the boundary of D_{ρ_1}. Consider now the function

$$w = \frac{1}{r} f(\rho_1 z) = F(z), \; F(0) = 0$$

with the following properties:

1) F renders a mapping of the circle $|z| < 1$ on to the region Δ containing the circle $|w| < 1$;

2) Δ has on the boundary a point with modulus 1;

3) the characteristic $p(z)$ of the transformation satisfies the condition $1 \le p(z) \le 1 + \varepsilon$;

4) the image of the ring γ: $(1 - \varepsilon_1)\dfrac{1}{n} < |z| < \dfrac{1}{n}$ contains the points $w_1' = w_1/r$, $w_2' = w_2/r$ such that

$$\left|\frac{w_2'}{w_1'}\right| > k. \tag{5.12}$$

Denote by $w = \Phi(z)$, $\Phi(0) = 0$ a function which maps the circle $|z| < 1$ conformally on to the region Δ. Let γ' be the image of the ring γ: $(1 - \varepsilon)/n < |z| < 1/n$ rendered by this transformation; by a well-known property of conformal transformations, there exists then a number n_0, $n_0 > 0$, which depends only on k and the number ε_1^0 such that for any points of γ' for $n \ge n_0$ and $\varepsilon_1 \le \varepsilon_1^0$

$$\left|\frac{\omega_2}{\omega_1}\right| < \frac{1+k}{2} < k; \tag{5.13}$$

on the other hand, one has

$$|\omega_1| > \frac{1-\varepsilon_1}{n}, \quad |\omega_2| > \frac{1-\varepsilon_1}{n}. \tag{5.14}$$

Now fix a number $n \geq n_0$ and select numbers ε and ε_1 so small that

1) $n\rho > 1$, which is possible because $\lim_{\varepsilon, \varepsilon_1 \to 0} \rho = 0$;

2) the distance between any pairs of points of the regions γ and γ' will be smaller than $[(k-1)/(k+1)]/2n$, which is possible by Theorem 5.2.

However, by (5.13) and (5.14), we have for these values of ε and ε_1 and for two arbitrary points ω_1' and ω_2' that $|\omega_2'/\omega_1'| < k$, and, in particular, $|w_2'/w_1'| < 1$, which contradicts the inequality (5.12). Thus, the theorem is proved.

5.2. The simplest class of linear systems

We will consider here the simplest class of linear elliptic systems of equations with first order partial derivatives, the so-called Beltrami systems, to which reduces the classical problem of conformal mapping of a surface. Let a manifold of homothetic ellipses with centre at a point z of the region D be linked to every such point; denote the ratio of the large semi-axes of these ellipses to the small semi-axes by $p(z)$, and the angle between the large axis and the x-axis by $\theta(z)$ (this function is only defined at those points where $p(z) > 1$). We will call the quantities p and θ the characteristics of the ellipses. The equation of the ellipses with characteristics p and θ can be rewritten in the form

$$\gamma(X-x)^2 - 2\beta(X-x)(Y-y) + \alpha(Y-y)^2 = ph^2, \quad (5.15)$$

where h is the small semi-axis and

$$\alpha = p\cos^2\theta + \frac{1}{p}\sin^2\theta,$$

$$\beta = \left(p - \frac{1}{p}\right)\sin\theta\cos\theta, \quad \alpha\gamma - \beta^2 \equiv 1. \quad (5.16)$$

$$\gamma = p\sin^2\theta + \frac{1}{p}\cos^2\theta,$$

The condition that the principal linear part of the transformation $w = f(z)$ at the point z is to map the ellipses of a given manifold on to circles can be written in the form of the system

$$\begin{aligned} \frac{\partial u}{\partial x} &= \beta \frac{\partial v}{\partial x} + \gamma \frac{\partial v}{\partial y}, \\ -\frac{\partial u}{\partial y} &= \alpha \frac{\partial v}{\partial x} + \beta \frac{\partial v}{\partial y}. \end{aligned} \tag{5.17}$$

We will confine attention to the case when the distribution of the ellipses is continuous in the region D, i.e., when the functions α, β, γ are continuous in this region and the solution $u + iv$ is continuously differentiable. However, within the framework of the theory under consideration, one can study arbitrary, measurable distributions of characteristics and generalized solutions of the system; such a generalization has been the subject of publications by a number of authors, the most definitive results in this direction being due to B. Boiarskii, L. Bers and L. Ahlfors.

The system (5.17) can also be rewritten in complex form by means of the formal derivative

$$\frac{\partial f}{\partial \bar{z}} = - e^{2i\theta} \frac{p - 1}{p + 1} \frac{\partial f}{\partial z}. \tag{5.18}$$

We will say that a function $f(z) = u + iv$ which satisfies this equation [or what is the same thing, the system (5.17)] is almost analytic of the class (p, θ).

5.2.1. Invariance with respect to conformal mappings. It is not difficult to show that the problem of the study of the entire class of almost analytic functions may be reduced to the theory of analytic functions by the aid of single-valued quasi-conformal transformations. In fact, if $w = f(z)$ is almost analytic in a region D and Δ is the region of the values of f, then, independently of the choice of the function $F(w)$, truly analytic in Δ, the function $F[(f(z)]$ will be almost analytic in D, where this function will belong to the same class (p, θ). In particular, the problem of the quasi-conformal mapping corresponding to the system (5.17) of an arbitrary simply-connected region D on to an arbitrary simply-connected region Δ reduces to the quasi-conformal mapping of D on to any definite region (for example, the unit circle) and the conformal mapping of this region (the circle) on to Δ. This statement is a straight consequence of the geometric definition of quasi-conformal transformations: the transformation f maps an ellipse with characteristics p, θ on to a circle, and the transformation F a circle on to a circle; consequently, the mapping $F(f)$ transforms the ellipse with characteristics p, θ into a circle.

The property of being almost analytic is likewise invariant to a conformal transformation of the plane of the independent variable: if the region of definition of the function $f(z)$ is to map conformally on to a new region D_1 of the z_1-plane, then in the region D_1 the ellipse (p, θ_1) will correspond to the ellipse with characteristics (p, θ), where the characteristic p is preserved and the characteristic θ becomes the new characteristic θ_1 which is equal to $\theta + \vartheta$, where ϑ is the angle of rotation of the conformal transformation.

Next, we will compare the quasi-conformal mappings $w = f(z)$ of the unit circle $|z| < 1$ on to the unit circle $|w| < 1$ which correspond to the system (5.17); without loss of generality, we can assume that $f(0) = 0$, $f(1) = 1$.

5.2.2. Stability of conformal mappings. We have already noted in the preceding section that quasi-conformal mappings are stable with respect to variations of the characteristics. This result is a simple consequence of Theorem 5.2:

THEOREM 5.4. *Let* $w = f(z)$ *and* $w = \tilde{f}(z)$, $f(0) = \tilde{f}(0) = 0$, $f(1) = \tilde{f}(1) = 1$ *be mappings of the circle* $|z| < 1$ *on to the circle* $|w| < 1$ *which are quasi-conformal with characteristics* p, θ *and* $\tilde{p}$, $\tilde{\theta}$, *respectively, where everywhere in the circle*

$$|\tilde{\alpha}(z) - \alpha(z)| < \varepsilon, \ |\tilde{\beta}(z) - \beta(z)| < \varepsilon, \ |\tilde{\gamma}(z) - \gamma(z)| < \varepsilon; \quad (5.19)$$

then everywhere in the circle

$$|\tilde{f}(z) - f(z)| < \lambda_1(\varepsilon), \quad (5.20)$$

where $\lambda_1(\varepsilon)$ *is a universal function which tends to zero for* $\varepsilon \to 0$.

In fact, by (5.19), in the case of a transformation

$$\zeta = f(z),$$

ellipses with characteristics $\tilde{p}$, $\tilde{\theta}$ become ellipses with characteristics $\tilde{\tilde{p}}$, $\tilde{\tilde{\theta}}$, where $\tilde{\tilde{p}} \leq 1 + \varepsilon$ everywhere inside the circle $|\zeta| < 1$.

1. Now let the function

$$w = F(\zeta), \; F(0) = 0, \; F(1) = 1$$

map the circle $|\zeta| < 1$ on to the circle $|w| < 1$ quasi-conformally with characteristics $\tilde{\tilde{p}}$, $\tilde{\tilde{\theta}}$. By Theorem 5.2, we have

$$F(\zeta) = \zeta + \lambda(\zeta, \varepsilon), \; |\lambda(\zeta, \varepsilon)| < \lambda_1(\varepsilon).$$

However,

$$\tilde{f}(z) = F[f(z)] = f(z) + \lambda(f, \varepsilon),$$

and hence

$$|\tilde{f}(z) - f(z)| < \lambda_1(\varepsilon),$$

and the theorem is proved.

As a supplement to this theorem, we present the following important consequence of Theorem 5.3:

THEOREM 5.5. *Let there be given in the circle* $|z| < 1$ *a sequence of characteristics* p_n, θ_n *such that* $\alpha_n(z)$, $\beta_n(z)$, $\gamma_n(z)$ *are continuous and converge uniformly in every closed part of the circle to* $\alpha(z)$, $\beta(z)$, $\gamma(z)$, *respectively. Then, if one can construct for every n a transformation*

$$w = f_n(z), \; f_n(0) = 0, \; f_n(1) = 1$$

of the circle $|z| < 1$ *on to the circle* $|w| < 1$ *which is quasi-conformal with characteristics* $p_n(z)$, $\theta_n(z)$, *the sequence* $f_n(z)$ *converges in* $|z| < 1$ *and the limiting function* $f(z) = \lim_{n\to\infty} f_n(z)$ *maps the circle* $|z| < 1$ *on to the circle* $|w| < 1$ *quasi-conformally with characteristics* $p(z)$, $\theta(z)$.

From this result follows immediately the existence theorem of a quasi-conformal mapping for any continuous pair of characteristics $p(z)$ and $\theta(z)$, if this theorem is true for the case when $p(z)$ and $\theta(z)$ are polynomials of step functions, i.e., functions which assume constant values in every square of some pattern of lines drawn on the circle $|z| < 1$.

5.2.3. Condition of smoothness of a transformation. The following question arises naturally: does the smoothness of a quasi-conformal mapping with given characteristics p and θ follow directly from the existence theorem, i.e., do the functions $u(x, y)$ and $v(x, y)$ have partial derivatives or continuous partial derivatives? It is not difficult to construct examples of contінous p and θ for which u and v will not have derivatives at isolated points or on certain manifolds of points. In this case, the mapping must be considered as a general solution of the system (5.17). We will now show that, if we impose on p and θ some additional limitations, the general solution will be automatically an ordinary solution of the system. [1]

We will begin by making Theorem 5.3 more precise for the case when the characteristics of the transformation satisfy Hölder conditions. We have the following

LEMMA 5.1. *Let there be given the quasi-conformal mapping*

$$w = f(z), \; f(0) = 0, \; f(1) = 1$$

of the circle $|z| < 1$ *on to the circle* $|w| < 1$ *the characteristic* $p(z)$ *of which satisfies the conditions*

$$\begin{gathered} 1 \leq p(z) \leq 1 + k|z|^{\nu}, \; 0 < \nu \leq 1, \\ 1 \leq p(z) \leq 1 + \varepsilon; \end{gathered} \tag{5.21}$$

then there exists at the point $z = 0$ *the maximal extension*

$$\lambda(0) = \overline{\lim_{|z|\to 0}} \; |f(z)|/|z|,$$

where

$$|\lambda(0) - 1| < k_1 \varepsilon \log \frac{1}{\varepsilon} \tag{5.22}$$

with k_1 *some constant.*

---

[1] This result was obtained by another method by B. V. Shabat, 1945.

For the purpose of the proof we will derive an estimate for the area $S(r)$ of the image of the circle $|z| < r$, given in Lemma 1.1. In the notation of this Lemma, we have

$$dS \geq \frac{2S\,dr}{rp}. \tag{5.23}$$

Using (5.21), we will improve this inequality setting in it

$$p = 1 + k|z|_\nu \ \text{ for } \ k|z|^\nu \leq \varepsilon,$$

$$p = 1 + \varepsilon \qquad \text{for } \ k|z|^\nu > \varepsilon.$$

Denote by $r_1 = (\varepsilon/k)^{1/\nu}$ the value of $|z|$ for which $k|z|^\nu = \varepsilon$ and, assuming r_0 to be small compared with r_1, integrate (5.23) from r_0 to 1, retaining in the computations only the principal terms. Thus

$$\log\frac{\pi}{S(r_0)} \geq 2\int_{r_0}^{1}\frac{dr}{rp} \geq 2\int_{r_0}^{r_1}\frac{dr}{r(1+kr^\nu)} + 2\int_{r_1}^{1}\frac{dr}{r(1+\varepsilon)} =$$

$$= 2\log\frac{r_1}{r_0} - \frac{2}{\nu}\log\frac{1+kr_1^\nu}{1+kr_0^\nu} + \frac{2}{1+\varepsilon}\log\frac{1}{r_1},$$

or, after raising to a power of e,

$$\frac{\pi}{S(r_0)} \geq \left(\frac{r_1}{r_0}\right)^2 \frac{1}{r_1^{2/(1+\varepsilon)}}\left(\frac{1+kr_0^\nu}{1+kr_1^\nu}\right)^{2/\nu}.$$

Replacing r_1 by its expression in terms of ε, noting that

$$\left(\frac{\varepsilon}{k}\right)^{-(1/\nu)[2\varepsilon/(1+\varepsilon)]} \approx 1 + \frac{2}{\nu}\varepsilon\log\frac{1}{\varepsilon}$$

and neglecting small higher order terms, we obtain, finally,

$$S(r_0) \leq \pi r_0^2\left(1 + \frac{2}{\nu}\varepsilon\log\frac{1}{\varepsilon}\right).$$

However, since we have $p = 1$ for $z = 0$, we find $S(|z|) \approx$

$\approx \pi|f(z)|^2$, i.e.,

$$\frac{|f(z)|}{|z|} \leq 1 + \frac{\varepsilon}{\nu} \log \frac{1}{\varepsilon},$$

whence follows the required upper bound for $\lambda(0)$.

In order to derive a lower bound for $\lambda(0)$, it is sufficient to investigate the inverse of the transformation f which, by virtue of Theorem 5.3, will also satisfy the conditions of the lemma (with the same ν and different k). Hence the lemma has been proved.

On the basis of this lemma, it is not difficult to prove a theorem concerning closeness of first order of quasi-conformal mappings corresponding to neighbouring systems, as well as a theorem relating to the smoothness of the solutions when the characteristics of the transformation fullfill a Hölder condition.

We will start with a few remarks on notation: let $|f'(z)|$ and $\arg f'(z)$ with

$$f'(z) = |f'(z)| e^{i \arg f'(z)}$$

denote the extension and rotation corresponding to the transformation $w = f(z)$ in the direction of the large axis of the characteristic ellipse, so that

$$|f'(z)|^2 = \left(\frac{\partial u}{\partial r}\right)^2 + \left(\frac{\partial v}{\partial r}\right)^2,$$

$$\arg f'(z) = \text{arc tan} \frac{\dfrac{\partial v}{\partial r}}{\dfrac{\partial u}{\partial r}},$$

where $\partial/\partial r$ indicates differentiation in the direction of the major axis of the characteristic ellipse.

THEOREM 5.6. *Let the functions*

$$w = f_1(z),\ f_1(0) = 0,\ f_1(1) = 1,$$
$$w = f_2(z),\ f_2(0) = 0,\ f_2(1) = 1$$

map the circle $|z| < 1$ *on to the circle* $|w| < 1$ *in a quasi-conformal manner with characteristics* p_1, θ_1 *and* p_2, θ_2, *respectively. If these characteristics lie close to each other and satisfy a Hölder condition, i.e.,*

$$|\alpha_2(z) - \alpha_1(z)| < \varepsilon, \ |\beta_2(z) - \beta_1(z)| < \varepsilon, \ |\gamma_2(z) - \gamma_1(z)| < \varepsilon,$$

$$|\alpha_i(z+h) - \alpha_i(z)|, \cdots, |\gamma_i(z+h) - \gamma_i(z)| < k|h|^\nu \quad (i = 1, 2),$$

then we have everywhere in the circle $|z| < 1$

$$|f_2'(z) - f_1'(z)| < M\varepsilon \log \frac{1}{\varepsilon}, \tag{5.24}$$

where the constant M *depends only on the constants* k *and* ν *in the Hölder conditions.*

THEOREM 5.7. *In the conditions of Theorem* 5.6, *each of the functions* $f_1(z)$, $f_2(z)$ *satisfy the inequality*

$$|f'(z+h) - f'(z)| < N|h|^\nu \log \frac{1}{h}, \tag{5.25}$$

where N *is a constant which depends only on* k *and* ν. *In addition, if*

$$1 \leq p(z) \leq 1 + \varepsilon,$$

then

$$|f'(z) - 1| < M\varepsilon \log \frac{1}{\varepsilon}.$$

We will prove Theorems 5.6 and 5.7 simultaneously. First of all, we establish in Theorem 5.6 the closeness of the extensions:

$$\big||f_2'(z)| - |f_1'(z)|\big| < M\varepsilon \log \frac{1}{\varepsilon};$$

it is not difficult to derive from this estimate the corresponding part of the estimate in Theorem 5.7:

$$\big||f'(z+h)| - |f'(z)|\big| < N|h|^\nu \log \frac{1}{|h|}.$$

Next, on the basis of these estimates, we prove the closeness of the arguments of f_1' and f_2' and derive from this result the second part of Theorem 5.7:

$$|\arg f'(z+h) - \arg f'(z)| < N|h|^\nu \log \frac{1}{|h|}.$$

For the purpose of the proof of the theorems, we perform certain preliminary reductions:

1) It is sufficient to study the case when the points z and $f(z)$ belong to the circles $|z| < \frac{1}{2}$, $|w| < \frac{1}{2}$. In fact, pursuing the method used earlier, we can extend the mapping to the circles $|z| < 2$, $|w| < 2$, and then consider the function $w = (\frac{1}{2})f(2z) = F(z)$, where the behaviour of $F(z)$ in the circle $|z| < \frac{1}{2}$ will correspond to that of f in the circle $|z| < 1$ (the constants ε and ν in Theorem 5.6 remaining the same, while k becomes $k2^\nu$).

2) It is sufficient to study the case when $p(z) \equiv 1$, i.e., when the first transformation is conformal. In fact, for the transformation

$$\zeta = f_1(z),$$

the ellipses with characteristics $p_2(z)$ and $\theta_2(z)$ become ellipses with characteristics $p(\zeta)$ and $\theta(\zeta)$ where, obviously,

$$p(\zeta) \leq 1 + \varepsilon,\ |\alpha(\zeta+h) - \alpha(\zeta)|,\ \cdots,\ |\gamma(\zeta+h) - \gamma(\zeta)| < k|h|^\nu. \tag{5.26}$$

Now let

$$w = F(\zeta),\ F(0) = 0,\ F(1) = 1$$

denote a transformation of the circle $|\zeta| < 1$ on to the circle $|w| < 1$ which is quasi-conformal with characteristics $p(\zeta)$ and $\theta(\zeta)$; obviously,

$$f_2(z) = F[f_1(z)].$$

However, then the required closeness of f_2' and f_1' will follow from the closeness of $F'(\zeta)$ to unity.

Noting this we find an estimate for $|F'(\zeta_0)| - 1$ for the conditions (5.26) and $|\zeta_0| < \frac{1}{2}$, $|w_0| = F(\zeta_0)| < \frac{1}{2}$. We execute an affine transformation of the ζ-plane such that an ellipse with characteristics $p(\zeta_0)$, $\theta(\zeta_0)$ becomes a circle; then the circle $|\zeta| < 1$ becomes an ellipse which differs from a circle by less than ε. This ellipse we transform conformally on to the unit circle with a corresponding normalization; the function which is the inverse to the superposition of these two transformations will be denoted by

$$\zeta = \varphi(z), \ \varphi(0) = \zeta_0, \ \varphi(1) = 1.$$

We will also denote by

$$\omega = \psi(w), \ \psi(w_0) = 0, \ \psi(1) = 1$$

a fractionally linear transformation of the circle $|w| < 1$ on to the circle $|\omega| < 1$ and consider the superposition

$$\omega = \psi\{F[\varphi(z)]\} = G(z), \ G(0) = 0, \ G(1) = 1.$$

This is a quasi-conformal mapping which satisfies the conditions of the lemma proved above, and hence

$$\big||G'(0)| - 1\big| < k\varepsilon \log \frac{1}{\varepsilon}.$$

However, $G'(0) = \psi'(\omega_0)F'(\zeta_0)\varphi'(0)$, where as a consequence of the properties of conformal and affine transformations

$$\big||\psi'(\omega_0)\varphi'(0)| - 1\big| \leq K\big||w_0| - |\zeta_0|\big| + \varepsilon$$

with K some constant. Substituting this result into the preceding inequality, we find

$$\big||F'(\zeta_0)| - 1\big| < M_1\varepsilon \log \frac{1}{\varepsilon} + K_1\big||w_0| - |\zeta_0|\big|, \qquad (5.27)$$

where M_1 and K_1 are constants. [1] Letting $h(|\zeta_0|) = |w_0| - |\zeta_0|$, we can rewrite (5.27) in the form

$$\frac{dh}{dr} \leq M_1\varepsilon \log \frac{1}{\varepsilon} + K_1 h,$$

[1] If we use here the theorem of P. P. Belinskii by which

$$\big||w_0| - |\zeta_0|\big| < 18\varepsilon,$$

the required estimate is obtained directly.

whence, since $h(0) = 0$, we find

$$h(r) \le \frac{M_1}{K_1} \varepsilon \log \frac{1}{\varepsilon} (e^{K_1 r} - 1).$$

Substituting for $h(r)$ into (5.27) this estimate, we obtain the required inequality

$$\big||F'(\zeta_0)| - 1\big| < M\varepsilon \log \frac{1}{\varepsilon},$$

and the first part of Theorem 5.6 has been proved.

Next, we will prove the first part of Theorem 5.7. For this purpose, we transform conformally with

$$z = \varphi(\zeta), \quad \varphi(z_0 + h) = z_0, \quad \varphi(1) = 1,$$

the circle $|\zeta| < 1$ on to the circle $|z| < 1$. Obviously, for $|h| < \frac{1}{4}$, we have

$$|\varphi'(\zeta) - 1| < 2|h|$$

for all ζ such that $|\zeta| < 1$, whence we conclude that for our transformations the system of ellipses with characteristics $p(\zeta)$ and $\theta(\zeta)$ becomes a system of ellipses with characteristics $p_1(z)$ and $\theta_1(z)$, where $p_1(z) = p(\zeta)$,

$$|\alpha_1(z) - \alpha(\zeta)|, \cdots. \; |\gamma_1(z) - \gamma(\zeta)| < 2|h|.$$

Thus, to obtain in the limit the accuracy required, we need only obtain bounds for the difference: $|f_1'(z_0)| - |f'(z_0)|$, where $f_1(z)$ brings about a quasi-conformal mapping of $|z| < 1$ on to the circle $|w| < 1$ with characteristics $p_1(z)$ and $\theta_1(z)$. However,

$$|\alpha_1(z) - \alpha(z)| = |\alpha(\zeta) - \alpha(z)| < k|\zeta - z|^\nu < k|h|^\nu,$$

and analogously

$$|\beta_1(z) - \beta(z)|, \; |\gamma_1(z) - \gamma(z)| < k|h|^\nu.$$

Further, by the conditions of the theorem, all the characteristics p, p_1, θ, θ_1 satisfy a Hölder condition with index ν; but then, by Theorem 5.6, we obtain for $|f_1'(z)| - |f'(z)|$, and hence also for $|f'(z + h)| - |f'(z)|$ the required bound

$$\big||f'(z + h)| - |f'(z)|\big| < k|h|^\nu \log \frac{1}{|h|}.$$

We note immediately that this method may also be applied to derive an estimate for $\arg f'(z+h) - \arg f'(z)$, when the second part of Theorem 5.6 will be proved, i.e., an estimate of $\arg f_2'(z) - \arg f_1'(z)$ will be derived.

In addition, one may set $f_1(z) \equiv z$, as this has been done above without violating generality, and thus reduce the entire task to an estimate of $\arg F'(z)$, where

$$w = F(z), \ F(0) = 0, \ F(1) = 1$$

maps the circle $|z| < 1$ on to the circle $|w| < 1$ and has characteristics $p(z)$ and $\theta(z)$ which satisfy the conditions

$$p(z) \leq 1 + \varepsilon, \ |\alpha(z+h) - \alpha(z)|, \cdots, |\gamma(z+h) - \gamma(z)| < k|h|^{\nu}.$$

In order to derive the estimate required, we draw through the point z_0, $|z_0| < \frac{1}{2}$ at which $\arg F'(z)$ is to be estimated an integral curve of the equation

$$\frac{dy}{dx} = \tan \theta(z).$$

This curve L divides the unit circle $|z| < 1$ into two parts; denote by D that part the boundary of which contains the point $z = 1$. We map D on to itself by use of the function

$$\zeta = \varphi(z), \ \varphi(z_0) = z_0, \ \varphi(1) = 1$$

in a quasi-conformal manner with characteristics $p(z)$ and $\theta(z)$.

By the first part of Theorem 5.6 and the condition that the curve L has at every point a tangent which satisfies a Hölder condition, we conclude that

$$\big|1 - |\varphi'(z)|\big| < M\varepsilon \log \frac{1}{\varepsilon},$$

$$\big||\varphi'(z+h)| - |\varphi'(z)|\big| < N|h|^{\nu} \log \frac{1}{|h|}.$$

Now, let $z = \psi(\zeta)$ be the inverse of the function φ and construct the function

$$\Phi(\zeta) = F[\psi(\zeta)].$$

This function is analytic in D, where we have

$$\left|\log |\Phi'(\zeta)|\right| < k\varepsilon \log \frac{1}{\varepsilon},$$

$$\left|\log |\Phi'(\zeta + h)| - \log |\Phi'(\zeta)|\right| < K |h|^{\nu} \log \frac{1}{|h|};$$

however, in that case, we find by a well known theorem relating to conjugate harmonic functions and by the condition $\varphi(1) = 1$ that

$$|\arg \Phi'(\zeta)| < R\varepsilon \log \frac{1}{\varepsilon}.$$

Besides,

$$|\arg \psi'(\zeta)| < \varepsilon^{\nu} \log \frac{1}{\varepsilon},$$

and hence

$$|\arg F'(z)| \leq |\arg \Phi'(\zeta)| + |\arg \psi'(\zeta)| <$$

$$< R\varepsilon \log \frac{1}{\varepsilon} + \varepsilon^{\nu} \log \frac{1}{\varepsilon} < A\varepsilon^{\nu} \log \frac{1}{\varepsilon}.$$

Thus, the theorem has been proved in its entirety.

5.2.4. Application to arbitrary linear systems. The above properties of quasi-conformal mappings corresponding to the simplest linear system (5.17) can be employed for the elucidation of analogous properties of quasi-conformal mappings, corresponding to arbitrary homogeneous systems of the elliptic type. We will present here one such theorem which will be required later on.

THEOREM 5.8. *Let $w = f(z) = u + iv$ be a quasi-conformal mapping of the region D corresponding to the elliptic system*

$$\frac{\partial u}{\partial y} = a_1 \frac{\partial u}{\partial x} + a_2 \frac{\partial v}{\partial x}, \qquad \frac{\partial v}{\partial y} = b_1 \frac{\partial u}{\partial x} + b_2 \frac{\partial v}{\partial x}, \tag{5.28}$$

where a_i and b_i are given functions of x, y, u, v which satisfy in D a Hölder condition and the condition of ellipticity

$$a_2b_1 + \left(\frac{a_1 - b_2}{2}\right)^2 < 0. \tag{5.29}$$

If in D the function $f(z)$ is bounded, $|f(z)| < m$, or its real part $|u| < m$, then in any region D_1, $\bar{D}_1 \subset D$, all its partial derivatives $\partial u/\partial x, \cdots, \partial v/\partial y$ will be bounded, i.e.,

$$\left|\frac{\partial u}{\partial x}\right| < Nm, \cdots, \left|\frac{\partial v}{\partial y}\right| < Nm, \tag{5.30}$$

where the number N depends on D, the closeness of D_1 to D and on the properties of the coefficients a, b, but not on the form of the function f.

Proof. The function $f(z)$ maps the region D on to some region Δ, where every infinitesimal circle of Δ corresponds to an infinitesimal ellipse with bounded characteristic p. We will map D on to the circle $|\zeta| < 1$ by means of the function

$$z = \varphi(\zeta)$$

in a quasi-conformal manner with characteristics $p(z)$, $\theta(z)$. The function

$$w = F(\zeta) = f[\varphi(\zeta)]$$

will be analytic and bounded in $|\zeta| < 1$; hence its derivative $F'(\zeta)$ will be bounded inside the circle in the sense inferred in the theorem, $|F'(\zeta)| < A \cdot m$. However, the derivative of the function $f(z)$ in any direction ν is equal to

$$f'_\nu(z) = \frac{F'(\zeta)}{\varphi'_\nu(\zeta)};$$

consequently, the task reduces to the derivation of an estimate for $\varphi'_\nu(\zeta)$ or for the derivative of the inverse of this function. Since the characteristic p of the transformation is bounded, the functions u and v satisfy in D_1 a Hölder condition; however, if we substitute in the coefficients a_k, b_k of the system (5.28) for u and v their expressions in terms of x and y from $w = f(z)$, the corresponding coefficients of the linear system will satisfy a Hölder condition. It follows from this that the mapping $z = \varphi(\zeta)$ has characteristics which satisfy a Hölder condition.

Thus, using Theorem 5.7, we find that $\varphi_\nu'(\zeta)$ exists, satisfies a Hölder condition and is non-zero, i.e., its modulus is bounded for all points of D_1 from above and below.

NOTE. Remaining in the same range of ideas and employing the concept of an arbitrary system, it may be shown that the theorem proved can be generalized in the following directions. If we assume, in addition, that the coefficients of the system (5.28) have derivatives with respect to all four arguments and all these derivatives satisfy a Hölder condition, estimates of the type (5.30) also hold true for the second derivatives of the functions u and v.

5.2.5 Existence theorem. For the solutions of many problems arising in connection with applications of quasi-conformal mappings great significance attaches to the general theorem of the existence and uniqueness of the quasi-conformal transformation of one region on to another for given characteristics p and θ. By virtue of the above invariance of quasi-conformal mappings with respect to conformal transformations, it is sufficient to study the mapping of the circle $|z| < 1$ on to the circle $|w| < 1$. We will now formulate the fundamental

THEOREM 5.9. *Let there be given in the circle* $|z| \leq 1$ *a continuous distribution of characteristics* $p(z)$ *and* $\theta(z)$. *There exists one and only one transformation*

$$w = f(z), \; f(0) = 0, \; f(1) = 1,$$

of the circle $|z| < 1$ *on to* $|w| < 1$ *for which at every point* z, $|z| < 1$, *an infinitesimal ellipse* $\mathscr{E}$ *with characteristics* $p(z)$ *and* $\theta(z)$ *becomes an infinitesimal circle.*

If it is assumed, in addition, that the characteristics p *and* θ *satisfy a Hölder condition*

$$|\alpha(z+h) - \alpha(z)| < k|h|^\nu, \cdots, |\gamma(z+h) - \gamma(z)| < k|h|^\nu,$$

the transformation $f(z)$ *is differentiable and the "derivative"* $f'(z)$

satisfies the condition

$$|f'(z+h) - f'(z)| < k|h|^\nu \log \frac{1}{|h|}.$$

We will begin with the proof of the uniqueness theorem. Assume that there exist two different mappings of a circle on to a circle with the same characteristics p and θ:

$$w = f_1(z), \quad w = f_2(z)$$

and with the same normalization $f_1(0) = f_2(0) = 0$, $f_1(1) = = f_2(1) = 1$. Then the system of equations

$$w = f_1(z), \quad \zeta = f_2(z)$$

establish a homomorphic correspondence between the circles $|w| < 1$ and $|\zeta| < 1$. This correspondence will transform any infinitesimal circle in the ζ-plane into an infinitesimal circle of the w-plane; however, by a theorem of D. E. Men'shov, this correspondence must be conformal [1]. By virtue of the adopted normalization, we find $w \equiv \zeta$, i.e., $f_2(z) \equiv f_1(z)$. Thus, the uniqueness theorem has been proved.

The existence theorem will be proved in three parts:

I. We establish first the existence of a single valued transformation with given characteristics $p(z)$ and $\theta(z)$ in a sufficiently small neighbourhood of the point z_0 which is exact or approximate for any complete class of characteristics p and θ.

In the capacity of an approximate transformation, one may select the affine transformation corresponding to the characteristics $p(z_0)$ and $\theta(z_0)$.

As exact solution, one can take the classical solution of the corresponding system of differential equations for the case when the coefficients are analytic functions of x and y; in this case, the required solution is obtained by evaluation of its Taylor series expansion. We note immediately that, since we are dealing

[1] If it is assumed that the mappings are continuously differentiable, conformality is a consequence of a classical theorem by Goursat. The proof under the assumptions of the theorem is due to Men'shov and requires delicate functional analytical methods.

with a single-valued transformation of a small region (for example, a circle or a square) on to such a region, one can construct, with the aid of a supplementary conformal transformation, a quasi-conformal mapping, with the same characteristics, of the small region on to an arbitrary region.

II. Next, we will prove a principle for the joining of solutions. Knowing single-valued mappings for regions with a common smooth boundary, we can construct a mapping of a region consisting of two given regions and their common boundary.

The patching principle is readily established either by a method of the type of Schwarz's alternating algorithm or with the help of the following

LEMMA 5.2. *Let there be given a real analytic function of the real variable x:*

$$x' = \varphi(x), \ \varphi(0) = 0, \ \varphi(1) = 1$$

such that for $0 \leq x \leq 1$, $\varphi'(x) > 0$. *Then one can construct two analytic functions*

$$f_1(z, h) \text{ and } f_2(z, h), \ z = x + iy, \ h > 0$$

such that:

a) f_1 *is regular and single-valued in the rectangle* $0 < x < 1$, $0 < y < h$ *and* f_2 *is regular and single-valued in the rectangle* $0 < x < 1$, $-h < y < 0$;

b) f_1 *and* f_2 *achieve "patching" of the upper and lower parts of the segment* $0 < x < 1$ *in accordance with the rule* $x' = \varphi(x)$, *i.e.,*

$$f_1(x, h) = f_2[\varphi(x), h];$$

c) *the function*

$$f = \begin{cases} f_1 \text{ for } y > 0 \\ f_2 \text{ for } y < 0 \end{cases}$$

is single-valued.

The lemma will be proved for sufficiently small h. Let

$$f_1(z, h) = \varphi(z), \ f_2(z, h) = z.$$

By virtue of this definition, f_1 and f_2 will satisfy the condition

b) and, by virtue of the analyticity of φ and the condition $\varphi' > 0$ for sufficiently small h, f_1 and f_2 will also satisfy the conditions a) and c).

We will assume that the lemma is correct for $h = h_0$ and will prove that it will also be true for $h = 2h_0$. Let D_1 and D_2 denote the images of the rectangles $0 < x < 1$, $0 < y < h_0$ and $0 < x < 1$, $-h_0 < y < 0$, and Γ the image of the segment $0 < x < 1$ of the x-axis for the transformations $f_1(z, h_0)$ and $f_2(z, h_0)$; let D denote the region formed by D_1, D_2 and Γ. We will map the region D conformally with the aid of the function $w = g(\zeta)$ on to the rectangle Δh_0: $0 < u < 1$, $-h_0 < v < h_0$, so that the points $\zeta_1 = f_1(1, h_0)$, $\zeta_2 = f_1(0, h_0)$, $\zeta_3 = f_2(0, -h_0)$, $\zeta_4 = f_2(1, -h_0)$ become the vertices of the rectangle Δh_0.

Let D_1' and D_2' denote the images of the regions D_1 and D_2 rendered by the transformation $w = g(\zeta)$ and consider the functions

$$w = g[f_1(z, h_0)] = f_1(z, 2h_0), \quad w = g[f_2(z, h_0)] = f_2(z, 2h_0).$$

It is readily seen that these are the functions sought. In fact, the function $f_1(z, 2h_0)$ renders a conformal mapping of the rectangle $0 < x < 1$, $0 < y < h_0$ on to the region D_1' and then the segment $0 < x < 1$, $y = h_0$, parallel to the x-axis, corresponds to the segment $0 < u < 1$, $v = h_0$, parallel to the u-axis. However, in that case, by the reflection principle of Schwarz, the function f_1 can be continued analytically into the rectangle $0 < x < 1$, $0 < y < 2h_0$. By virtue of these considerations, f_2 will be analytic and single-valued in the rectangle $0 < x < 1$, $-2h_0 < y < 0$. In this way, the property a) has been proved for the functions f_1 and f_2; the properties b) and c) follow from those of $f_1(x, h_0)$ and $f_2(x, h_0)$ and from the construction of $f_1(x, 2h_0)$, $f_2(x, 2h_0)$. Thus the lemma has been proved.

We will now show how the above patching principle follows from this lemma. Let the simply connected regions D_1 and D_2 with a common (sufficiently) smooth boundary segment Γ admit the quasi-conformal mappings

$$w = f_1(z), \quad w = f_2(z)$$

with characteristics p and θ. We will map the images of D_1 and D_2 in the w-plane conformally on to the lower and upper halves of the square $(0, -\frac{1}{2})$, $(1, -\frac{1}{2})$, $(1, \frac{1}{2})$, $(0, \frac{1}{2})$ of the $\zeta = \xi + i\eta$ plane; we will dispose of the arbitrariness in the mappings in such a way that the arc Γ becomes for both transformations the segment $(0, 1)$ of the ξ-axis. In this manner, we obtain quasi-conformal mappings with characteristics p and θ of the regions D_1 and D_2 on to the unit square, where every point a of the curve Γ will correspond to two points, say ξ and ξ', of the ξ-axis. Applying the lemma above, we can subdivide by the curve γ the unit square in the ω-plane into two regions Δ_1 and Δ_2 such that on conformal transformations of the halves of the square in the ζ-plane on to Δ_1 and Δ_2 the points ξ and ξ' go over into the same point of the curve γ. Thus, we obtain a quasi-conformal mapping with characteristics p and θ of the region $D_1 + \Gamma + D_2$ on to the region $\Delta_1 + \gamma + \Delta_2$, where this transformation is continuous and single-valued. This proves the patching principle.

III. It follows from I and II that one can construct for any continuous distribution of characteristics $p(z)$ and $\theta(z)$ a quasi-conformal transformation with characteristics arbitrarily close to $p(z)$ and $\theta(z)$; whatever may be the value of the number $n \geq 1$, one can construct polynomials or step functions $p_n(z)$ and $\theta_n(z)$ such that one will have in the circle $|z| \leq 1$

$$|\alpha_n(z) - \alpha(z)| < \frac{1}{n}, \cdots, |\gamma_n(z) - \gamma(z)| < \frac{1}{n}.$$

By I and II, there exists a transformation of the circle $|z| < 1$ on to the circle $|w| < 1$

$$w = f_n(z), \; f_n(0) = 0, \; f_n(1) = 1$$

with characteristics $p_n(z)$ and $\theta_n(z)$. It remains to prove that $f_n(z)$ tends for $n \to \infty$ to the transformation required.

From the boundedness of $p_n(z)$ follows the compactness of the manifold $\{f_n(z)\}$; from Theorem 5.5 follows that every limiting function will render a mapping with characteristics $p(z)$ and $\theta(z)$. However, since such a mapping is unique, it follows that the sequence $\{f_n(z)\}$ has a unique limiting function, i.e., that $f_n(z)$ converges uniformly to the transformation required.

Thus, the existence theorem has been proved completely.

In conclusion, we will show that in the case when the characteristics $p(z)$ and $\theta(z)$ satisfy a Hölder condition the derivative $f'(z)$ exists and also satisfies a Hölder condition. In fact, in this case the characteristics $p(z)$ and $\theta(z)$ can be approximated by the characteristics $p_n(z)$ and $\theta_n(z)$ (analytic or step functions) which also satisfy a Hölder condition with the same index[1]:

$$|\alpha_n(z + h) - \alpha_n(z)| < k|h|^\nu,$$
$$\cdots\cdots\cdots\cdots$$
$$|\gamma_n(z + h) - \gamma_n(z)| < k|h|^\nu;$$

however, by Theorem 5.7, we have for all approximate mappings $f_n(z)$

$$|f_n'(z + h) - f_n'(z)| < K|h|^\nu \log \frac{1}{|h|},$$

where the constant K does not depend on n. Thus, in this case, not only the manifold of the transformations, but also of their derivatives is compact, i.e., the limiting transformation $f(z)$ will have at every point a derivative $f'(z)$ which satisfies a Hölder condition.

---

[1] For the case when the characteristics are step functions, the inequalities will be satisfied for $h \gg 1/n$.

CHAPTER VI

THE SIMPLEST CLASSES OF NON-LINEAR SYSTEMS

We will present here in detail one class of strongly elliptic non-linear systems, i.e., of systems of the form

$$\begin{aligned} \Phi_1\left(\frac{\partial u}{\partial x}, \frac{\partial u}{\partial y}, \frac{\partial v}{\partial x}, \frac{\partial v}{\partial y}\right) &= 0, \\ \Phi_2\left(\frac{\partial u}{\partial x}, \frac{\partial u}{\partial y}, \frac{\partial v}{\partial x}, \frac{\partial v}{\partial y}\right) &= 0, \end{aligned} \tag{6.1}$$

which involve explicitly neither the independent variables x and y nor the unknown functions u and v. In terms of characteristics, the system (6.1) assumes the form

$$\begin{aligned} W &= F_1(V, \alpha), \\ \theta &= F_2(V, \alpha), \end{aligned} \tag{6.2}$$

where by virtue of the strong ellipticity of the system the functions F_1 and F_2 have the properties

$$\begin{gathered} \frac{\partial W}{\partial V} \geq k_0 > 0, \\ 0 < k_0 < \theta < 2\pi - k_0 \end{gathered} \tag{6.3}$$

for all values of α and V within the region of definition, k_0 being some constant.

We will show that the fundamental variational principles of the theory of conformal mappings extends to the quasi-conformal transformations corresponding to strongly elliptic systems of the type (6.1). We will start with generalizations of the maximum principles for the modulus and the argument of the derivative.

6.1. Maximum Principle

By the results of Chapter V, the functions $P = \log V$ and α satisfy a quasi-linear system of equations of the elliptic type, the derivative system

$$\begin{aligned} \frac{\partial P}{\partial v} &= a_1 \frac{\partial P}{\partial u} + a_2 \frac{\partial \alpha}{\partial u}, \\ \frac{\partial \alpha}{\partial v} &= b_1 \frac{\partial P}{\partial u} + b_2 \frac{\partial \alpha}{\partial u}. \end{aligned} \tag{6.4}$$

It follows from the ellipticity and homogeneity of the system (6.4) that any one of its solutions P, α renders a quasi-conformal mapping with bounded distortion of the region of definition of the functions P, α in the (u, v)-plane on to some simply-connected Riemann surface located above the (P, α)-plane.

Hence follows directly the classical

THEOREM 6.1 (Maximum Principle). *Let $w = f(z)$ achieve a quasi-conformal mapping, corresponding to the strongly elliptic system* (6.1), *of a simply-connected region D of the (x, y)-plane on to some region Δ of the (u, v)-plane. Then the functions P and α can assume their extreme values only on the boundary of D.*

With a view towards applications to gas dynamics, we will concentrate our attention on the mapping of the strip $D(\Gamma_0, \Gamma)$: $y_0(x) < y < Y(x)$ on to the strip Δ: $0 < v < h$; we will assume that the functions $y_0(x)$ and $Y(x)$ are twice differentiable and that their second derivatives satisfy a Hölder condition:

$$|y_0''(x + \Delta x) - y_0''(x)| < k|\Delta x|^\nu,$$
$$|Y''(x + \Delta x) - Y''(x)| < k|\Delta x|^\nu.$$

Let

$$w = f(z), \quad f(\pm \infty) = \pm \infty,$$

denote a function which renders a quasi-conformal transformation of the strip D on to the strip Δ. In this process, let the curve

$$y = y(x, v)$$

of the strip D become the straight line $v =$ const. We will call this curve a stream line or an equipotential line. Now, two new functions will be introduced: the density of flow line

$$R = \frac{\partial y}{\partial v}$$

and the gradient of flow line

$$\tau = \tan \alpha = \frac{\partial y}{\partial x}.$$

It is readily seen that in this notation the system (6.4) assumes the form

$$\frac{\partial R}{\partial x} = \frac{\partial \tau}{\partial v}, \qquad \frac{\partial R}{\partial v} = a \frac{\partial \tau}{\partial x} + b \frac{\partial \tau}{\partial v}, \tag{6.5}$$

where a and b are known functions of R and τ. Under the conditions of strong ellipticity the derivative system of the system (6.1) will likewise be strongly elliptic [1], i.e,, for all values of the arguments, the following inequality will hold true:

$$-a - \tfrac{1}{4}b^2 \geq k > 0.$$

As before, it follows from this that the functions R and α can attain values only on the boundary of D.

6.2. The principle of Schwarz-Lindelöf

For the class of transformations under consideration, we will now establish the second fundamental principle, i.e., the principle of Schwarz-Lindelöf. For the purpose of the ensuing considerations, we will assume from the start that the functions, which

[1] It has been assumed here that one is not dealing with an exceptional case (cf. 4.3).

render the corresponding transformations, exist and have two first order partial derivatives which satisfy a Hölder condition. In addition, we will assume that, if the "boundary" functions $y_0(x)$ and $Y(x)$ tend to finite limits as $x \to \pm\infty$, also $y(x, v)$ tends for $x \to \pm\infty$ and fixed v to finite limits, and these limiting values depend only on the corresponding limits y_0, Y and on v.

Side by side with the strip $D(\Gamma_0, \Gamma)$: $y_0(x) < y < Y(x)$ consider a strip $D(\Gamma_0, \tilde{\Gamma})$: $y_0(x) < y < \tilde{Y}(x)$ which lies close to it and let $y = y(x, v)$, $y = \tilde{y}(x, v)$ be flow lines for quasi-conformal mappings of these strips on to the strip $0 < v < h$.

We pose now the problem of estimates of the variations of the flow line in dependence on variations of the boundary Γ. We will first prove the following

LEMMA 6.1. *Let*

$$M(v) = \max_{|x|<\infty} [\tilde{y}(x, v) - y(x, v)]$$

and assume that the functions y_0, Y and $\tilde{Y}$ tend to finite limits as $|x| \to \infty$. Then

$$\frac{d^2M}{dv^2} \geq A \frac{dM}{dv}, \tag{6.6}$$

where A is some constant which depends on the smoothness of the flow lines $y = y(x, v)$ and $\tilde{y} = \tilde{y}(x, v)$.

We will employ for the proof the second relation (6.5):

$$\frac{\partial^2 y}{\partial v^2} = a \frac{\partial^2 y}{\partial x^2} + b \frac{\partial^2 y}{\partial x \partial v},$$

$$\frac{\partial^2 \tilde{y}}{\partial v^2} = \tilde{a} \frac{\partial^2 \tilde{y}}{\partial x^2} + \tilde{b} \frac{\partial^2 \tilde{y}}{\partial x \partial v},$$

where $\tilde{a}$ and $\tilde{b}$ are the values of a and b corresponding to the changed transformation. Letting

$$z = z(x, v) = \tilde{y}(x, v) - y(x, v),$$

we find

$$\frac{\partial^2 z}{\partial v^2} = a\,\frac{\partial^2 z}{\partial x^2} + b\,\frac{\partial^2 z}{\partial x \partial v} + \left[\frac{\partial a}{\partial R}\,\frac{\partial^2 \tilde{y}}{\partial x^2} + \frac{\partial b}{\partial R}\,\frac{\partial^2 \tilde{y}}{\partial x \partial v}\right]\frac{\partial z}{\partial v} +$$
$$+ \left[\frac{\partial a}{\partial \tau}\,\frac{\partial^2 \tilde{y}}{\partial x^2} + \frac{\partial b}{\partial \tau}\,\frac{\partial^2 \tilde{y}}{\partial x \partial v}\right]\frac{\partial z}{\partial x} =$$
$$= a\,\frac{\partial^2 z}{\partial x^2} + b\,\frac{\partial^2 z}{\partial x \partial v} + A\,\frac{\partial z}{\partial v} + B\,\frac{\partial z}{\partial x}. \tag{6.7}$$

Taking note of this result, consider in the (x, v)-plane points P and P' at which the functions $z(x, v)$ and $z(x, v + dv)$, respectively, assume maximum values as functions of x [1]. The angle α between the direction PP' and the x-axis is obviously determined by the relation

$$\frac{\partial^2 z}{\partial x^2}\cos\alpha + \frac{\partial^2 z}{\partial x \partial v}\sin\alpha = 0. \tag{6.8}$$

We will now find the derivative of the function $\partial z/\partial v$ in the direction α

$$\frac{d}{dt}\,\frac{\partial z}{\partial v} = \frac{\partial^2 z}{\partial x \partial v}\cos\alpha + \frac{\partial^2 z}{\partial v^2}\sin\alpha$$

or, using (6.7) and (6.8),

$$\frac{d}{dt}\,\frac{\partial z}{\partial v} = a\,\frac{\partial^2 z}{\partial x^2}\sin\alpha + (b\sin\alpha + \cos\alpha)\,\frac{\partial^2 z}{\partial x \partial v} +$$
$$+ \left(A\,\frac{\partial z}{\partial v} + B\,\frac{\partial z}{\partial x}\right)\sin\alpha = -\frac{1}{\sin\alpha}\,\frac{\partial^2 z}{\partial x^2}[-a\sin^2\alpha + \cos^2\alpha +$$
$$+ b\sin\alpha\cos\alpha] + \left(A\,\frac{\partial z}{\partial v} + B\,\frac{\partial z}{\partial x}\right)\sin\alpha. \tag{6.9}$$

By virtue of the strong ellipticity of the fundamental system, the square bracket is positive for all α; besides, we will have at the point P, as a point of a maximum of the function $z(x, v)$, that

$$\frac{\partial^2 z}{\partial x^2} \le 0,\quad \frac{\partial z}{\partial x} = 0,\quad \frac{\partial z}{\partial v} = \frac{dM}{dv},\quad \frac{1}{\sin\alpha}\,\frac{d}{dt}\,\frac{dM}{dv} = \frac{d^2 M}{dv^2}.$$

---

[1] Such points A and A' exist in a bounded part of the strip, because by virtue of the above assumptions we have $z \to 0$ for $|x| \to \infty$.

Thus, the formula (6.9) assumes at P the form

$$\frac{d^2M}{dv^2} = -\frac{1}{\sin^2\alpha}\frac{\partial^2}{\partial x^2}[-a\sin^2\alpha + \cos^2\alpha + b\sin\alpha\cos\alpha] + A\frac{dM}{dv},$$

whence follows the required inequality

$$\frac{d^2M}{dv^2} \geq A\frac{dM}{dv}.$$

NOTE. It has been assumed for the proof that α is non-zero; however, for the case $\alpha = 0$, it is sufficient to take instead of dv a finite increment Δv and, without modifying the manipulations, one will arrive at the above result.

Next, the fundamental theorem will be proved.

THEOREM 6.2 (Principle of Schwarz-Lindelöf). *Under the condition of the lemma proved above, if*

$$\tilde{Y}(x) \geq Y(x),$$

one has for any v, $0 < v < h$,

$$\tilde{y}(x, v) \geq y(x, v); \tag{6.10}$$

further, at every point of Γ_0,

$$\frac{\partial\tilde{y}}{\partial v} \geq \frac{\partial y}{\partial v}, \tag{6.11}$$

at the points of Γ *where* $\tilde{Y}(x) - Y(x)$ *attains a maximum,*

$$\frac{\partial\tilde{y}}{\partial v} \geq \frac{\partial y}{\partial v}, \tag{6.12}$$

and at points common to Γ *and* $\tilde{\Gamma}$ [*where* $\tilde{Y}(x) = Y(x)$]

$$\frac{\partial\tilde{y}}{\partial v} \leq \frac{\partial y}{\partial v}. \tag{6.13}$$

The equality signs in (6.10)–(6.13) *apply only in the case when* $\tilde{Y}(x) \equiv Y(x)$.

Proof. We can limit consideration to the case when

$$\lim_{x\to\pm\infty} y_0'(x) = \lim_{x\to\pm\infty} Y'(x) = 0, \quad \lim_{x\to\pm\infty} [\tilde{Y}(x) - Y(x)] = 0.$$

As in Lemma 6.1, let $M(v)$ and $m(v)$ denote the maximum and minimum of $\tilde{y}(x, v) - y(x, v)$, respectively, for fixed v, By the conditions of the theorem, we will have $m(0) = M(0) = 0$.

For the case of a minimum, Lemma 6.1 assumes the form

$$\frac{d^2m}{dv^2} \leq A \frac{dm}{dv}, \tag{6.14}$$

where A is some constant. The integral of the differential equation

$$\frac{d^2m}{dv^2} = A \frac{dm}{dv}$$

for the initial conditions

$$m(0) = 0, \ m'(0) < 0$$

is obviously negative for all $v > 0$. On the basis of this observation and the inequality (6.14), it may be asserted that, if $dm/dv|_{v=0} < 0$, then all the more $m(v) < 0$ for all v, $0 < v \leq h$. However, this contradicts the fact that, by assumption, $\tilde{Y}(x) \geq Y(x)$.

Thus, we will have $dm/dv|_{v=0} \geq 0$. If we had $dm/dv|_{v=0} = 0$, then we would have everywhere for $v = 0$

$$\tilde{y} = y, \quad \frac{\partial \tilde{y}}{\partial v} = \frac{\partial y}{\partial v}.$$

By virtue of the uniqueness of the solution of the Cauchy problem, we find that in that case $\tilde{y}(x, v) \equiv y(x, v)$ for all v, $0 \leq v \leq h$, despite of the conditions of the theorem. Classical results ensure the uniqueness of the solution of this problem under the condition that the left-hand sides of the system (6.1) are analytic functions. Without the assumption of analyticity, this result can be obtained either by construction of a majorant for the variations $\delta(\partial y/\partial v)$ or by transition to a study of a derivative system which in the variables R, α becomes a linear system, and by use of a result due to M. M. Lavrent'ev (1957) relating to the uniqueness of the solutions of Cauchy's problem for such systems.

Thus, it has been shown that $dm/dv|_{v=0} > 0$, i.e., that at every point of Γ_0

$$\frac{\partial \tilde{y}}{\partial v} > \frac{\partial y}{\partial v},$$

and the inequality (6.11) has been proved.

If one starts out from points common to Γ and $\tilde{\Gamma}$, one can obtain by completely analogous reasoning the inequality (6.13). The inequality (6.12) is readily obtained from the inequalities (6.11) and (6.13). In fact, the variation of $\partial y/\partial v$ at the point x_0 of a maximum of $\tilde{Y}(x) - Y(x)$ can be obtained by superposition of the variations of $\partial y/\partial v$ under the following two variations of Γ:

1) the curve Γ is translated in the direction of the y-axis by an amount $\tilde{Y}(x_0) - Y(x_0)$;
2) the resulting curve is transferred into the curve $\tilde{\Gamma}$.

By virtue of (6.11) and (6.13), $\partial y/\partial v$ will decrease for each of these variations, and hence we arrive at the required inequality (6.12).

We still have to prove the inequality (6.10). In contradiction, assume that for some value of the parameter v, $0 < v < h$, the curve $y = \tilde{y}(x, v)$ has a point with a smaller ordinate than the curve $y = y(x, v)$. However, since we have, by (6.11), for sufficiently small v that $\tilde{y}(x, v) > y(x, v)$, we find consequently a value of the parameter v: $v = v_0$, for which

$$\tilde{y}(x, v_0) \geq y(x, v_0),$$

where for some value of x: $x = x_0$, we have $\tilde{y}(x_0, v_0) = y(x_0, v_0)$.

When applied to the strips

$$y_0(x) < y < y(x, v_0), \quad y_0(x) < y < \tilde{y}(x, v_0),$$

the inequality (6.13) renders at the point $[x_0, y(x_0, v_0)]$

$$\frac{\partial \tilde{y}}{\partial v} \leq \frac{\partial y}{\partial v}. \tag{6.15}$$

On the other hand, the inequality (6.11), applied to the strips

$$y(x, v_0) < y < y(x, h), \quad y(x, v_0) < y < \tilde{y}(x, h)$$

renders at the same point

$$\frac{\partial \hat{y}}{\partial v} > \frac{\partial y}{\partial v},$$

where $\hat{y}$ refers to the function, mapping $y(x, v_0) < y < \tilde{y}(x, h)$ on to the strip $v_0 < v < h$, and application of (6.13) to the strip

$$y(x, v_0) < y < \tilde{y}(x, h), \quad \tilde{y}(x, v_0) < y < \tilde{y}(x, h)$$

renders at the same point

$$\frac{\partial \hat{y}}{\partial v} \leq \frac{\partial \tilde{y}}{\partial v}.$$

Thus, at the point $[x_0, y(x_0, v_0)]$, we have

$$\frac{\partial \tilde{y}}{\partial v} > \frac{\partial y}{\partial v},$$

despite the inequality (6.15). The theorem has now been proved in its entirety.

All properties of conformal mappings, resting on the principle of Schwarz and Lindelöf, can be extended to the class of quasi-conformal transformations under consideration. As an example of the application of this principle, we will prove the uniqueness theorem of a mapping.

THEOREM 6.3 (uniqueness). *A twice differentiable, quasi-conformal transformation, corresponding to a strongly elliptic system* (6.1) *of a strip* $D(\Gamma_0, \Gamma)$ *on to a strip* $0 < v < h$ *under the condition of correspondence of infinitely distant points is determined uniquely apart from a translation in the* (u, v)*-plane.*

Proof. In contradiction, we assume that two different mappings exist which do not reduce to each other by a translation; then there will exist for the same $y_0(x)$ and $Y(x)$ two equi-potential lines $y = y(x, v_0)$ and $y = \tilde{y}(x, v_0)$ for the same value of the parameter $v = v_0$, $0 < v_0 < h$. We will consider a point x_0 at which the difference $\tilde{y}(x, v_0) - y(x, v_0)$ attains a maximum value. On the basis of the principle of Schwarz-Lindelöf, we may now compare the values of $\partial y/\partial v$ and $\partial \tilde{y}/\partial v$ at the points $[x_0, y(x_0, v_0)]$ and $[x_0, \tilde{y}(x_0, v_0)]$, investigating these expressions for the strips $y_0(x) < y < y(x, v_0)$, $y_0(x) < < y < \tilde{y}(x, v_0)$, as well as for the strips $y(x, v_0) < y < Y(x)$, $\tilde{y}(x, v_0) < y < Y(x)$. If we obtain for one set of strips $\partial y/\partial v > > \partial \tilde{y}/\partial v$, then we find for the other $\partial y/\partial v < \partial \tilde{y}/\partial v$. This contradiction proves completely the uniqueness theorem.

6.3. Quantitative estimates

The maximum principle and the principle of Schwarz-Lindelöf permit to construct for actual, strongly elliptic systems qualitative and quantitative bounds for different important functionals related to corresponding quasi-conformal mappings. At this stage, we will present without detailed proofs a number of results in this direction.

THEOREM 6.4. *Let the boundaries Γ_0 and Γ of the region $D(\Gamma_0, \Gamma)$ satisfy the conditions*

$$\begin{gathered} k_0 \leq Y(x) - y_0(x) \leq k, \\ |y_0'(x)| \leq k', \quad |Y'(x)| \leq k', \\ |y_0''(x)| \leq k'', \quad |Y''(x)| \leq k''; \end{gathered} \tag{6.16}$$

then in the region D

$$0 < m < \frac{\partial y(x, v)}{\partial v} < M, \tag{6.17}$$

where the constants m and M depend only on the constants k and the functions F_1, F_2, and not on the shape of the region D.

By virtue of the maximum principle, it is sufficient to prove the theorem for the points of Γ_0 and Γ. The methods for the derivation of upper and lower bounds of $R = \partial y/\partial v$ are the same and one can restrict oneself to obtaining an estimate of the lower bound for points of Γ. By the principle of Schwarz-Lindelöf, it is sufficient to determine R for any "majorant" region $D(\gamma_0, \gamma)$, where γ_0 and γ must have the following properties: γ touches Γ and lies nowhere below Γ; γ_0 does not lie below Γ_0. By this method one can find the estimate (6.17).

For various applications, as also in the simplest case of conformal mappings, an important role is played by estimates for the variations of $\partial y/\partial v$ in dependence on variations of Γ. We will consider immediately the derivation of such estimates in the case of mappings of strips, close to the unit strip

$$0 < y < 1,$$

on to the strip $0 < v < 1$ of the (u, v)-plane. Without restricting generality, we will assume for this purpose that $y_0(x) \equiv 0$, $Y(x) > 0$, $\delta Y \geq 0$, where $\lim \delta Y(x) = 0$ for $x \to \pm\infty$ with δY tending to zero as quickly as we may require.

First of all, it is not difficult to see that for $Y(x) \equiv 1$, if

$$|\delta Y| \leq \varepsilon, \quad |\delta Y'| < \varepsilon, \quad |\delta Y''| < \varepsilon, \tag{6.18}$$

one has everywhere

$$|1 - R| = \left| 1 - \frac{\partial y(x, v)}{\partial v} \right| \leq k\varepsilon, \quad |\tau| \equiv \left| \frac{\partial y}{\partial x} \right| \leq k\varepsilon, \tag{6.19}$$

where the constant k does not depend on ε. If we assume, in addition, that $\delta Y \equiv 0$ for $x \geq 0$, we have for $x \geq 0$

$$|1 - R| < k\varepsilon e^{-x}, \quad |\tau| < k\varepsilon e^{-x}, \tag{6.20}$$

where the constant k depends on constants characterizing the smoothness of the functions F_1, F_2 and a constant, characterizing the strong ellipticity of the system under consideration. This result is obtained, if one notes that for $Y(x) \equiv 1$ the mapping will be affine with $y(x, v) \equiv v$.

We will still present certain formulae for the variations of $\partial y/\partial v$; for this purpose, we write down, first of all, the equation for the function $y(x, v)$

$$\frac{\partial^2 y}{\partial v^2} - a \frac{\partial^2 y}{\partial x^2} - b \frac{\partial^2 y}{\partial x \partial v} = 0. \tag{6.21}$$

If the curve $\Gamma : y = 1$ is replaced by $\tilde{\Gamma} : y = 1 + \varepsilon$, the new solution will be

$$y(x, v) = (1 + \varepsilon)v,$$

and we obtain for the variation of $\partial y/\partial v$

$$\delta \frac{\partial y}{\partial v} = \varepsilon.$$

Consider now several other particular cases of variations of δY. Let $\delta Y \geq 0$; we conclude from the principle of Schwarz-Lindelöf, proved above, that at a point where δY attains a maximum we will have

$$\delta \frac{\partial y}{\partial v} \geq \max \delta Y, \tag{6.22}$$

where the equality sign applies only in the case when $\delta Y \equiv \varepsilon$.

Now let $\delta Y > 0$ for $x_0 < x < x_0 + l$ and $\delta Y \equiv 0$ outside this interval and write

$$\sigma = \int_{x_0}^{x_0+l} \delta Y\, dx = \nu l \max \delta Y,$$

where ν is some fixed constant. Then we will have for points of the curve $\Gamma_0 : y = 0$

$$\delta \frac{\partial y}{\partial v} = \theta \sigma e^{-(x-x_0)}, \tag{6.23}$$

where $0 < \theta_0 < \theta < 1/\theta_0$ and θ_0 is some constant which depends only on ν and on the smoothness of δY.

Analogously, we will have for points of the straight line $\Gamma : y = 1$, located outside the segment $(x_0, x_0 + l)$,

$$\delta \frac{\partial y}{\partial v} = -\, \theta \sigma e^{-(x-x_0)}. \tag{6.24}$$

Hence one can obtain as a linear approximation a general estimate for the variations of the derivative at an arbitrary point x_0 of the x-axis and arbitrary sufficiently smooth variations of the boundary $y = 1$; in fact,

$$\delta \frac{\partial y}{\partial v} = \theta \int_{-\infty}^{+\infty} \delta Y \cdot e^{-|x-x_0|} dx. \tag{6.25}$$

From this representation for $\delta(\partial y/\partial v)$ one can obtain also an estimate for $\delta y(x, v)$. For example, we will have for $0 < v < \lambda < 1$

$$\delta y(x, v) > \theta v \sigma e^{-|x-x_0|}, \tag{6.26}$$

where θ is some constant which depends on v and λ.

In linear formulation, for sufficient closeness of Γ to the straight line $y \equiv 1$ and Γ_0 to the x-axis, all the properties enumerated above for the variations $\delta(\partial y/\partial v)$ and δy carry over to the strip $D(\Gamma_0, \Gamma)$. This transfer may be placed on a rigorous foundation, if the equation (6.21) is subjected to a variation for the transition from the solution $y(x, v)$ close to $y(x, v) \equiv v$ to an infinitely close solution.

For $z = \delta y(x, v)$, we obtain

$$\frac{\partial^2 z}{\partial v^2} - a_0 \frac{\partial^2 z}{\partial x^2} - b_0 \frac{\partial^2 z}{\partial x \partial v} = \left[\frac{\partial^2 y}{\partial x^2} \frac{\partial a}{\partial \tau} + \frac{\partial^2 y}{\partial x \partial v} \frac{\partial b}{\partial \tau} \right]_0 \frac{\partial z}{\partial x} + \\ + \left[\frac{\partial^2 y}{\partial x^2} \frac{\partial a}{\partial R} + \frac{\partial^2 y}{\partial x \partial v} \frac{\partial b}{\partial R} \right]_0 \frac{\partial z}{\partial v}, \tag{6.27}$$

where the subscript 0 relates to the values of the functions when y is replaced by its expression in the original solution $y(x, v)$. If we neglect the right-hand side of (6.27), then lower bounds for $z = \delta y(x, v)$ and its derivatives with respect to v are obtained from the selected simplest case. It is seen that these variations are of the order of the initial variations of the boundary conditions. These estimates must be made more exact by taking into consideration the right-hand side of (6.27); however, for small $\partial z/\partial x$ and $\partial z/\partial v$, we have there small coefficients, where the order of smallness of the coefficients is determined by the closeness of the basic solution to the linear solution $y = v$. For the purpose of a completely rigorous proof, one would have to obtain closer estimates for the second derivatives of $y(x, v)$ in dependence on the boundary values.

It appears to be more effective to obtain this proof by construction of majorants for $\delta[\partial y(x, v)/\partial v]$ within definite classes of curves Γ_0, Γ and within a specified class of strongly elliptic systems. We will dwell on several applications of these formulae and make a beginning with approximate formulae for R in the case of mappings of narrow strips. In this case, formulae analogous to those in Chapter I may be obtained.

In fact, let $D(\Gamma_0, \Gamma)$ be a "narrow" strip $y_0(x) < y < Y(x)$ in the sense that for small h

$$k_0 h < Y(x) - y_0(x) < k_1 h$$

and the first three derivatives of $y_0(x)$ and $Y(x)$ are bounded by certain constants:

$$|y_0'| < k', \ |Y'| < k', \ |y_0''| < k'', \ |Y''| < k'', \ |y_0'''| < k''', \ |Y'''| < k'''.$$

Also let $w = f(z) = u + iv$ be a quasi-conformal mapping of D on to the strip $0 < v < h$. Select an arbitrary point of the strip D and let ds denote the arc element of the equipotential line $y = y(x, v)$ of the transformation and dn the element of the normal to this line.

The derivative system (6.4) may be rewritten in the form

$$\frac{\partial P}{\partial n} = a_{11}\frac{\partial \alpha}{\partial s} + a_{12}\frac{\partial \alpha}{\partial n},$$
$$\frac{\partial P}{\partial s} = a_{21}\frac{\partial \alpha}{\partial s} + a_{22}\frac{\partial \alpha}{\partial n}. \tag{6.28}$$

The point selected above will be assumed to be a "median" point of the strip, for example, the point halving the segment of the normal to Γ_0, contained within D. Now we can determine along this normal all the characteristics of the transformation exactly apart from small quantities of second order compared with h. In fact, the geometric characteristic α can be determined directly, and the quantities $\partial\alpha/\partial s$ and $\partial\alpha/\partial n \approx \Delta\alpha/\Delta n$ are determined straight away apart from a quantity of order h.

Hence we find from the basic equation in terms of characteristics

$$W = F_1(V, \alpha) \approx \frac{\Delta n}{h}$$

the mean value $P_0 = \log V_0$. Substituting these quantities into the first equation of the system (6.28), we obtain the unknown distribution of P along the normal

$$P - P_0 = \left[a_{11} \frac{\partial \alpha}{\partial s} + a_{12} \frac{\partial \alpha}{\partial n} \right] t + \rho,$$

where t is the distance from the median point of the normal, $- \Delta n/2 < t < \Delta n/2$, and ρ is the remainder term. The variational formulae derived above permit to obtain estimates for ρ; obviously, these estimates will be of the same order as in the case of a conformal transformation.

In conclusion, we will still present a theorem relating to the same circle of questions.

THEOREM 6.5. *If the curves Γ_0 and Γ have a curvature k satisfying a Hölder condition, then in the case of a quasi-conformal transformation of the region $D(\Gamma_0, \Gamma)$ on to the strip $0 < v < 1$, which corresponds to the strongly elliptic system* (6.1), *the functions u and v have partial derivatives of second order which satisfy a Hölder condition.*

The theorems of Chapter II relating to the behaviour at points of not large inclination of Γ and at points where k attains a maximum extend to this class of quasi-conformal transformations.

6.4. Inductive proof of Lindelöf's principle

In view of the fundamental significance of Lindelöf's principle, we will still present one of its proofs based on the method of continual induction. In the following section, we will prove by this method another important result of the theory of quasi-conformal mappings, i.e., the theorem of the existence of a transformation of a region $D(\Gamma_0, \Gamma)$ on to the strip.

Our proof will be based on the fact that the principle is true for a strip $D(\Gamma_0, \Gamma)$ as close as one pleases (in the sense of second or third order closeness) to the unit strip $0 < y < 1$. For this case, the principles may apparently be proved by classical methods; one can also prove them by use of the approximate formulae obtained in the last section for narrow strips.

We require for the proof also the following

LEMMA 6.2. *Let the function $w = f(z)$ map the strip $D(\Gamma_0, \Gamma)$ quasi-conformally on to the strip $0 < v < h$, where $\Gamma_0 : y = y_0(x)$ and $\Gamma : y = Y(x)$ are such that $y_0(x)$ and $Y(x)$ have three bounded derivatives; also let $w = \tilde{f}(z)$ map the strip $D(\Gamma_0, \tilde{\Gamma})$, where $\tilde{\Gamma} : y = \tilde{Y}(x)$, on to the same strip $0 < v < h$ and let*

$$|y_0'| < k', \ |Y'| < k', \ |\tilde{Y}(x) - Y(x)| < \varepsilon. \tag{6.29}$$

Then there exists a function $\nu(\varepsilon)$, $\lim_{\varepsilon \to 0} \nu(\varepsilon) = 0$, which does not depend on $\tilde{Y}$ and is such that

$$\left| \tilde{y}'\left(x, \frac{h}{2}\right) - y'\left(x, \frac{h}{2}\right) \right| < \nu(\varepsilon). \tag{6.30}$$

Proof. In fact, by virtue of the maximum principle, we will have along Γ_0

$$|\delta R| < K\varepsilon;$$

on the other hand, we have along the same curve $\delta\tau = 0$. Thus, we have two mappings $\omega = R + i\tau$ and $\tilde{\omega} = \tilde{R} + i\tilde{\tau}$ both of which are bounded and have infinitely close Cauchy boundary conditions; however, then these transformations will also be close in the interior of the region.

On the basis of this lemma and the invariance of the solutions under linear transformations of all variables: $\xi = \lambda x$, $\eta = \lambda v$, $\zeta = \lambda y$, it is readily shown that the conclusion of the lemma remains true, if we replace the strip $0 < v < h$ by the strip $0 < v < h + \varepsilon'$ with ε' sufficiently small.

We will now proceed to the proof of Lindelöf's principle. For the proof of the principle in its general form, it will be sufficient to show that, if it is true for mappings on to a strip $0 < v < h$ of strips $D(\Gamma_0, \Gamma)$, where Γ_0 and Γ satisfy the conditions

$$|y_0'(x)| \leq k', \ |Y'(x)| \leq k', \ |y_0''(x)| \leq k'', \ |Y''(x)| \leq k'',$$
$$0 < k_0 h \leq Y(x) - y_0(x) \leq k_1 h, \qquad (6.31)$$
$$\lim_{x\to\pm\infty} y_0(x) = \lim_{x\to\pm\infty} y_0'(x) = \lim_{x\to\pm\infty} Y'(x) = 0, \ \lim_{x\to\pm\infty} Y(x) = h,$$

it will also be true for strips for which Γ_0 and Γ satisfy the same conditions, but with larger values of the constants k_0, k_1, k', k''. We note still that the variations may be considered to be infinitely small and that the values of the constants k_0 and k_1 are inessential, because for similarity transformations of the z- and w-planes in the same ratio the mapping remains in the former class of quasi-conformal transformations. For similarity transformations of one of the planes, we obtain again a quasi-conformal mapping, the only difference being that the variable V appears in the expressions for the functions F_1 and F_2 with a constant multiplier; thus, also in the case of these transformations, we do not leave the limits of the class of quasi-conformal mappings under consideration.

Thus, we will assume that for $k' \leq k_0'$ and $k'' \leq k_0''$ the principle is true.

We will now select for the number k_1' a fixed value so small that for the transformation of any region of the class of regions, satisfying the conditions (6.31), one would have for $k' < k_1'$

$$\left| \frac{\partial^2 y(x, h/2)}{\partial x^2} \right| < \tfrac{1}{2}k''. \qquad (6.32)$$

Such a value of k_1' exists by virtue of the lemma proved above. We replace now the class (6.31) for $k' \leq k_1'$, $k'' \leq k_0''$ by the same class for $k' \leq k_1'$, $k'' \leq k_0'' + \varepsilon$, where ε is some sufficiently small constant. By (6.32), we will have

$$\left| \frac{\partial^2 y(x, h/2)}{\partial x^2} \right| < \tfrac{1}{2}(k'' + \varepsilon) < k'',$$

whence, applying extensions to both planes z and w, we conclude that we can apply the principle to each of the strips

$$y_0(x) < y < y\left(x, \frac{h}{2}\right), \quad y\left(x, \frac{h}{2}\right) < y < Y(x) \tag{6.33}$$

(the curve $y = y(x, h/2)$ may not be close to either of the curves $y = y_0(x)$ and $y = Y(x)$, because, by Theorem 6.2, the quantity $|f'(z)|$ in our class of transformations is bounded from above and below).

We still have to prove that it follows from the condition $\delta Y(x) > 0$ that $\delta y(x, h/2) > 0$. In contrast, we will assume that this is not true and denote by x_0 the point at which $\delta y(x, h/2)$ attains a minimum; by assumption, one has

$$\delta y\left(x_0, \frac{h}{2}\right) \leq 0. \tag{6.34}$$

We apply now the principle of Schwarz-Lindelöf to the strips (6.33) and the varied strips $y_0(x) < y < \tilde{y}(x, h/2)$, $\tilde{y}(x, h/2) < < y < Y(x)$, respectively, at the points with abscissa x_0. By (6.34), we will have for the variations of the lower strips at the points

$$\left[x_0, y\left(x_0, \frac{h}{2}\right)\right] \text{ and } \left[x_0, \tilde{y}\left(x_0, \frac{h}{2}\right)\right]$$

the result

$$\delta \frac{\partial y}{\partial v} < 0,$$

and for the variations of the upper strips at the same points

$$\delta \frac{\partial y}{\partial v} > 0;$$

thus, a contradiction is obtained and the statement has been proved.

There remains to show that the constant k' bounding the gradients of Γ_0 and Γ can be increased. We introduce again into the consideration two strips: $y_0(x) < y < y(x, h/2)$ and $y(x, h/2) < y < Y(x)$. By virtue of the maximum principle for

τ, we will have

$$\left| \frac{\partial y(x, h/2)}{\partial x} \right| < k';$$

thus, for one of the strips: $y_0(x) < y < y(x, h/2)$ or $y(x, h/2) < < y < Y(x)$, our principle is true (i.e., for the first strip, if $Y(x)$ has been varied, for the second, if $y_0(x)$ has been varied). For the sake of definiteness, let it be true for the lower strip and $\delta Y > 0$. As before, we will consider a point x_0 at which $\tilde{y}(x, h/2) - y(x, h/2)$ attains a minimum and let, in contrast, $\tilde{y}(x_0, h/2) \leq y(x_0, h/2)$; however, by virtue of the inductive assumptions, we will have for the transformations of the upper strips on to the strip $h/2 < v < h$ at the points $[x_0, y(x_0, h/2)]$ and $[x_0, \tilde{y}(x_0, h/2)]$,

$$\delta \frac{\partial y}{\partial v} \leq 0,$$

while we will have at those points for the mappings of the lower strips on to the strip $0 < v < h/2$ under the condition $\tilde{y}(x, h/2) \not\equiv y(x, h/2)$

$$\delta \frac{\partial y}{\partial v} < 0,$$

which is impossible. The remaining case $\tilde{y}(x, h/2) \equiv y(x, h/2)$ is excluded, because in this case, by virtue of the uniqueness of the Cauchy problem, we would have that $\tilde{y}(x, v) \equiv y(x, v)$, i.e., $\tilde{Y}(x) \equiv Y(x)$.

6.5. The existence theorem

The most important result of the theory of quasi-conformal mappings is the generalization of Riemann's theorem regarding the existence of a conformal mapping of a given region on to another. It turns out to be possible to extend this theorem to mappings corresponding to a class of strongly elliptic systems. In fact, one has

THEOREM 6.6. *Irrespectively of what may be the strongly elliptic system* (6.1), *the two simply connected regions D and Δ,* *bounded by smooth curves C and Γ and the points* $z_0 \in D$, $w_0 \in \Delta$, $z_1 \in C$, $z_2 \in \Gamma$, *there exists one and only one quasi-conformal transformation*

$$w = f(z), \; w_0 = f(z_0), \; w_1 = f(z_1)$$

which maps D on to Δ and corresponds to the system (6.1).

We will prove this theorem by use of the same inductive method by which we proved Lindelöf's principle in the preceding section. We will restrict ourselves to the case, which is most important for applications, when the region D is a curvilinear strip $D(\Gamma_0, \Gamma)$ and Δ is the strip $0 < v < h$. We will assume that the theorem has been proved for regions $D(\Gamma_0, \Gamma)$, close, in the sense of second order closeness, to a rectilinear strip, i.e., for solutions close to the linear solution: $w = kz$; with regard to this statement, we can repeat what has been said in 6.4 with respect to Lindelöf's principle. We will only give the principles of the proof of the theorem in the general case.

First of all, we will state a more precise formulation of the problem. We will seek the solution of the Dirichlet problem

$$y(x, 0) = y_0(x), \; y(x, h) = Y(x)$$

for the equation of the elliptic type

$$\frac{\partial^2 y}{\partial v^2} - a \frac{\partial^2 y}{\partial x^2} - b \frac{\partial^2 y}{\partial x \partial v} = 0, \tag{6.35}$$

where a and b are given, three times differentiable functions of the variables $R = \partial y / \partial v$ and $\tau = \partial y / \partial x$.

The solution of this Dirichlet problem will be sought in the strip $0 < v < h$ for any equation (6.35) of some class E of equations and boundary conditions, satisfying certain restrictions. With respect to the class of equations, we will accept the following conditions:

1) the constant determining the strong ellipticity of the basic system, and likewise of the system of equations for R and τ, is fixed;

2) the moduli of all first, second and third order partial derivatives of the functions a and b are bounded by a constant k. With

respect to the boundary values $y_0(x)$ and $Y(x)$, we will introduce the following restrictions:

$$0 < y_0(x) < Y_1(x) \leq k_0, \ |Y(x) - y_0(x)| \leq k_1,$$
$$|y_0'(x)| \leq k', \ |Y'(x)| \leq k',$$
$$|y_0''(x)| \leq k'', \ |Y''(x)| < k''.$$

We will assume inductively that the solution of the problem under consideration exists, when the constants k, k' and k'' are sufficiently small or, what is the same thing, when these constants have certain values. We will prove that one can then construct also a solution for larger values of these constants; the theorem will then be proved in the general form.

First of all, we note for the purpose of the proof that the solutions of the equation (6.35) are invariant with respect to the linear transformation $\xi = \lambda x$, $\eta = \lambda v$, $\zeta = \lambda y$. We note also that for a similarity condensation of the (x, v)-plane in the ratio λ, y' and y'' decrease λ and λ^2 times, respectively. We conclude from these two properties that it will be sufficient for a complete proof of the theorem to show that for sufficiently small ε, ε' a solution will exist, if we replace the curve $Y(x)$ by the curve $Y(x) + \varepsilon$ and the constant h, determining the width of the strip, by $h + \varepsilon'$.

Denote by ν'' the upper bound of $\partial^2 y(x, v)/\partial x^2$ for the initial solution y

$$\left| \frac{\partial^2 y(x, v)}{\partial x^2} \right| \leq \nu''$$

and select a number n so large that

$$\frac{\nu''}{n} \leq \frac{k''}{2}. \tag{6.36}$$

Then subdivide the strip $0 < v < h$ into $2n$ parts

$$H_i : \frac{ih}{2n} < v < \frac{(i + 1)h}{2n}, \ i = 0, \cdots, 2n - 1$$

and select ε, ε' so small that, for variations of the values of

$y(x, 2ih/2n)$ and $y(x, 2(i+1)h/2n)$ of the function $y(x, v)$ on the boundaries of the strips $v_i = 2ih/2n < v < 2(i+1)h/2n = v_{i+1}$ by ε and for expansion of these strips by ε'/n the variation of $\partial^2 y/\partial x^2$ on the centre line of this strip, $v = (2i+1)h/2n$ will not exceed $k''/2$. Such values of ε and ε' can be found by virtue of the lemma in the preceding section.

We will now proceed to the construction of the solution. Since a change from h to $h + \varepsilon'$ does not cause any principal modification of the construction, we will confine ourselves to the case $\varepsilon' = 0$. In the strip $H_{2(n-1)} + H_{2n-1}$, we construct the solution y_{11} for boundary conditions $y(x, v_{n-1})$ and $y(x, h) + \varepsilon = Y(x) + \varepsilon$, in the strip $H_{2(n-2)} + H_{2n-3}$, the solution y_{12} for the boundary conditions $y_{11}(x, v_{n-2})$ and $y(x, v_{n-1})$, etc., up to the strip $H_0 + H_1$, where the solution is constructed for the given boundary values $y_0(x)$ and $y_{12n-1}(x, v_1)$. Subsequently, we revert to the strip $H_{2(n-1)} + H_{2n-1}$ and construct there a solution $y_{21}(x, v)$ for the boundary conditions $y_{12}(x, v_{n-1})$ and $Y(x) + \varepsilon$, etc.

By virtue of Lemma 6.2, in each boundary value problem, the second derivatives of the boundary values will not exceed $2v''$, i.e., after reduction of the strip H_i to the strip $0 < v < h$, the second derivatives will not exceed k''. Hence we have proved that all the functions of the sequence can be constructed on the basis of the earlier assumptions relating to the existence of the solution for sufficiently small k.

Thus, in the strip $0 < v < h$, we obtain the sequence of functions

$$y_{1i}(x, v), \; y_{2i}(x, v), \; \cdots, y_{mi}(x, v), \; \cdots$$

In each of the strips $v_i < v < v_{i+1}$, this sequence will decrease monotonically, by virtue of the principle of Schwarz-Lindelöf, and, consequently, it will converge to some function

$$y(x, v) = \lim_{m \to \infty} y_{mi}(x, v).$$

The limiting function has continuous derivatives and therefore, by virtue of the compactness of the class of quasi-conformal mappings corresponding to the derived system, it will satisfy the equation (6.35) and the varied boundary conditions.

If we note, in addition, that for all functions of the sequence $R_m = \partial y_{mi}/\partial v > 0$, then it will follow from this and the compactness that the constructed function $y = y(x, v)$ will render equipotential lines of a single-valued quasi-conformal transformation of the strip $y_0(x) < y < Y(x) + \varepsilon$ on to the strip $0 < v < h$.

Thus, the existence theorem has been proved in its entirety.

6.6. Generalizations

The existence theorem, just as certain general properties of the mapping functions, can be generalized to the case of an arbitrary strongly elliptic system. However, even in the case of linear systems, the principle of Schwarz-Lindelöf is preserved only in a weakened form. We will cite here one of the relevant theorems.

THEOREM 6.7. *Let there be given the strongly elliptic system which does not involve explicitly the unknown functions:*

$$W = F_1(x, y, V, \alpha),$$
$$\theta = F_2(x, y, V, \alpha),$$

where, as before, the functions F_1 *and* F_2 *possess all partial derivatives (of the first three orders). Further, let the functions* $w = f(z)$ *and* $w = \tilde{f}(z)$ *render quasi-conformal mappings of the regions* $D(\Gamma_0, \Gamma)$ *and* $D(\Gamma_0, \tilde{\Gamma})$, *respectively, on to the strip* $0 < v < h$ *in such a manner that the points at infinity correspond. If* Γ_0, Γ *and* $\tilde{\Gamma}$ *satisfy the conditions of the principle of Schwarz-Lindelöf in* 6.2 *and, in addition,* $D(\Gamma_0, \Gamma)$ *is a sufficiently narrow strip, i.e.,*

$$0 < Y(x) - y_0(x) < h,$$

where h *is sufficiently small, then all the conclusions of this principle apply.*

The constant h is determined by the constant, which characterizes the strong ellipticity of the system, and the constants limiting the derivatives of W and θ with respect to the independent variables.

The proof of the formulated theorem can be obtained by the method applied earlier to the proof of the analogous theorem in the case when F_1 and F_2 do not depend explicitly on x and y. The following reasoning will also lead to this result.

We will decrease h without bound, reducing in this process the strips $D(\Gamma_0, \Gamma)$ and $0 < v < h$ to strips of finite width, i.e., we will expand the (x, y)- and (u, v)-planes by similarity transformations in the ratio $1/h$; then the equations in characteristics will assume the form

$$W = F_1^*(x, y, V, \alpha) = F_1(hx, hy, V, \alpha),$$

$$\theta = F_2^*(x, y, V, \alpha) = F_2(hx, hy, V, \alpha).$$

As $h \to 0$, our transformations will tend to the transformation corresponding to the system

$$W = F_1(0, 0, V, \alpha),$$

$$\theta = F_2(0, 0, V, \alpha)$$

which no longer involves the coordinates x and y explicitly.

Assuming, in contrast, that one of the conclusions of the theorem is untrue and using the compactness of the class of mappings for $h \to 0$, we arrive at a contradiction to Theorem 6.3.

This method also permits to expand the theorems relating to the behaviour of V at the points of maximum inclination of Γ and at points of extreme curvature of Γ to the case of narrow strips.

6.7. Hydrodynamic applications

In the case of the equations of gas dynamics – in the plane case as well as in that of axial symmetry – the general theorems proved here lend themselves to obvious mechanical interpretations. We will begin with the plane case where the most complete results are available.

The plane case. As it has been stated above, the problem of the quasi-conformal mapping, corresponding to the system

$$\mu = f(\lambda), \quad \theta = \frac{\pi}{2}, \tag{6.37}$$

(where $\lambda = 1/V$ is the velocity and $\mu = 1/W$ is the transport) of the region $D(\Gamma_0, \Gamma)$ on to the strip $0 < v < H$ is equivalent (for a special function f) to the problem of the construction of a plane steady gas flow streaming between the walls Γ_0 and Γ with transport H. For subsonic flow ($\lambda < \lambda_s$), we have $f'(\lambda) > 0$, i.e., the system leads to a strongly elliptic system.

Now, let $\Gamma_0 : y = y_0(x)$ and $\Gamma : y = Y(x)$ satisfy the conditions

$$\begin{aligned} k_0 &\leq Y(x) - y_0(x) \leq k_1, \\ |y_0'(x)| &< k', \quad |Y'(x)| < k', \\ |y_0''(x)| &< k'', \quad |Y''(x)| < k'', \end{aligned} \tag{6.38}$$

then an H_0 may be found such that for any $H \leq H_0$ there exists a gas flow with transport H between the walls Γ_0 and Γ. For $H = H_0$, there will be a point on the boundary of the flow at which the velocity of the flow is equal to the velocity of sound.

We will now present a brief justification of this statement.

For this purpose, replace the function f by a function f_1 which is equal to f for $\lambda \leq \lambda_s - \varepsilon$ and which increases everywhere; extrapolating f_1 from the value $\lambda = \lambda_s - \varepsilon$ to the remaining values of λ, we proceed in such a manner that f_1 will satisfy all the conditions of the existence theorem and compactness. We will assume the number ε to be a constant which is as small as we please.

By the existence theorem proved earlier for the system

$$\mu = f_1(\lambda), \quad \theta = \frac{\pi}{2},$$

there exists a quasi-conformal mapping of $D(\Gamma_0, \Gamma)$ on to the strip $0 < v < H$ for any H, where for this transformation, by virtue of (6.38), the function λ will be bounded and its maximum will be obtained on the boundary of $D(\Gamma_0, \Gamma)$. By virtue of this result and the similarity principle, we will have for $H \to 0$ that $\max \lambda \to 0$. Thus, for H sufficiently small everywhere within the region $D(\Gamma_0, \Gamma)$, we will have for our transformation $\lambda < \lambda_s - \varepsilon$. This proves the existence of the flow for sufficiently small transport H. In addition, it is obvious that $\max \lambda$ will be a linear function of H; thus, we obtain the existence of the flow up to some H for which $\max \lambda \leq \lambda_s - \varepsilon$. Hence, by virtue of the arbitrariness of ε, a solution exists for all subsonic regimes. However, since inside D the velocity λ is always smaller than on the boundary (except for the trivial case $\lambda = \text{const.}$, when the flow region is a rectilinear strip), i.e.,

$$\lambda < k \max \lambda_b \leq \lambda_s,$$

where k depends only on the distance of the points from the boundary of D, it is not difficult to derive from this that for $\varepsilon \to 0$, $H \to H_0$, $\max \lambda \to \lambda_s$ the limiting flow exists. This proves completely our statement above.

The existence theorem of quasi-conformal transformations together with the last properties can be extended to the case of multiply connected regions; thus we arrive at the possibility of constructing gas flows around bodies with circulation. A theorem on the existence of flows around contours with circulation may be established for all subsonic regimes.

We will still consider briefly the problem of the construction of a flow around a body with break away. Since the principle of Schwarz-Lindelöf extends completely in its quantitative and qualitative forms to quasi-conformal mappings, corresponding to the equations of gas dynamics, as well as to the deductions made from it with regard to the behaviour of λ on the boundary of the mapped region, we conclude from this that the method developed above for the construction of stream line flows for an incompressible ideal fluid can also be extended to the case of an ideal gas. Reasoning as before in the case of ordinary flow, we can obtain the solution of this problem for all subsonic regimes.

The case of axial symmetry. In the case of a flow with axial symmetry, two additional difficulties are encountered:

1) along the axis of symmetry $y = 0$, the system of equations determining the motion

$$W = \frac{1}{y} f(V), \quad \theta = \frac{\pi}{2},$$

2) the principle of Schwarz-Lindelöf and its consequences apply only for sufficiently narrow strips located at a distance from the axis of symmetry.

For the problem of stream line flow (without discontinuities), both these circumstances are inessential: we can construct, in accordance with the general theory, a quasi-conformal mapping of a region $D(\Gamma_0, \Gamma)$

$$\Gamma_0 : y = \varepsilon$$
$$\Gamma_1 : y = Y(x) > h > \varepsilon$$

on to the strip $\varepsilon < v < H$, and will obtain the motion in the tube $0 < y < Y(x)$ (i.e., in the region bounded by the surface of revolution with cross-section $y = Y(x)$ by the limiting process $\varepsilon \to 0$.

As in the plane case, we establish the existence of the solution for all subsonic regimes. More exactly, the solution will exist for sufficiently small transport which may increase until somewhere in the flow the velocity becomes equal to the velocity of sound. As in the plane case, the passage through the velocity of sound requires an essentially new method.

In conclusion, we will still refer to problems of free stream lines and consider a very simple case.

Let there be given a curve $\Gamma_0 : y = y(x)$ with bounded gradient and curvature such that outside the segment $(0, 1)$

$$y_0(x) = a > 0,$$

and on the segment $(0, 1)$

$$y_0(x) \geq a, \; y_0(1) = y_0(0) = a.$$

It is required to find a curve $\Gamma : y = Y(x)$ such that for a quasi-conformal mapping of the region $D(\Gamma_0, \Gamma)$ on to the strip

$0 < v < H$, one will have along Γ

$$V = V_0 = \text{const.};$$

the quasi-conformal transformation is to correspond to the system of equations of gas dynamics in the case of symmetry.

It is clear from the earlier analysis that the solution of the problem exists and is unique, if the quantities V_0 and $\lambda = H/V_0$ are sufficiently small. In this case, the parameter λ characterizes the width of the strip $D(\Gamma_0, \Gamma)$: the smaller is λ, the closer the velocity of the flow will approach to the velocity of sound.

The problem of the flow around a body with a free stream line may be solved in this case by analogy with the simplest plane case of an ideal incompressible fluid; our method is applicable, if the region of flow in the region of the unknown free stream-line is sufficiently narrow for a definite subsonic regime.

REFERENCES

I. CONFORMAL TRANSFORMATIONS

Monographs

FIL'CHAKOV, P. F. The theory of filtration under hydro-technical structures. Vols. 1 and 2, Kiev, Izd-vo AN SSSR, 1959, 1960.

KOBER, H. Dictionary of conformal representations. Dover, New York, 1952.

KOPPENFELS, F. and STALLMAN, S. Die Praxis der konformen Abbildungen. Springer, Berlin, 1959.

LAVRENT'EV, M. A. Conformal transformations with applications to certain problems of mechanics. Moscow-Leningrad, Gostekhizdat, 1946.

LAVRENT'EV, M. A. and SHABAT, B. V. The methods of the theory of functions of a complex variable. 2nd edition, Moscow, Fizmatgiz, 1958.

NEHARI, Z. Conformal mapping. 1st edition, Mac Graw-Hill, New York, 1952.

Papers

BEZSONOV, P. A. and LAVRENT'EV, M. A. Sur l'existence de la dérivée limite. Bull. Soc. math. France, **58**, 1930, 175–198.

IVANILOV, IU. P., MOISEEV, N. N. and TER-KRIKOROV, A. M. On the asymptotic character of the formulae of M. A. Lavrent'ev, Dokl. AN SSSR, **123**, 1958, 231–234.

KRAVTCHENKO, J. Représentation conforme de Helmholtz. Théorie des sillages et des proues. J. Math. Pures Appl., **20**, 1941, 34–301.

LAVRENT'EV, M. A. On the theory of conformal transformations. Trudy Fiz.-Matem. In-ta AN SSSR, **5**, 1934, 159–246.

LAVRENT'EV, M. A. On certain boundary value problems in the theory of single-valued functions. Matem. Sbornik, **1** (43), 1936, 815–846.

LAVRENT'EV, M. A. On certain properties of single-valued functions with applications to the theory of jets. Matem. Sbornik **4** (46), 1938, 391–458.

LERAY, J. and WEINSTEIN, A. Sur un problème de réprésentation conforme posé par la théorie de Helmholtz. C. R. Paris, **198**, 1934, 430–432.

II. HYDRODYNAMIC APPLICATIONS

Monographs

BIRKHOFF, G. Hydrodynamics. Dover, New York. 1955.

BIRKHOFF, G. and ZARANTANELLO, E. H. Jets, waves and cavities, New York, 1960.

NEKRASOV, A. I. The exact theory of stationary waves on the surface of a heavy fluid. Moscow, Izd-vo AN SSSR, 1951.

SRETENSKII, L. N. The theory of wave motions of a fluid. Moscow-Leningrad, Gostekhizdat, 1936.

STOKER, J. J. Water waves. Interscience Publishers, New York, 1957.

Papers

FRIEDRICHS, K. O. and HYERS, D. H. The existence of solitary waves. Communications in Pure and Applied Mathematics, Vol. 7, 1954, 517–550.

GERBER, R. Sur une classe de solutions des équations du mouvement avec surface libre d'un liquide pesant. C. R. Acad. Sci., **242**, 1956, 1260–1262.

GERBER, R. Sur les solutions exactes des équations du mouvement avec surface libre d'un liquide pesant. (Thesis, Univ. de Grenoble, 1955), Journal de Mathématiques, 1955.

KELDYSH, M. V. and LAVRENT'EV, M. A. On the motion of a wing under the surface of a heavy fluid. Presentation at the Conference on Wave Resistance, 21.3.1936.

KELLER, J. The solitary wave and periodic in shallow water. Comm. Pure Appl. I, 1948, 323–328.

LAVRENT'EV, M. A. On certain properties of jet motion. Dokl. AN SSSR, **20**, 1938, 235–238.

LAVRENT'EV, M. A. On the theory of jet motion. Dokl. AN SSSR, **20**, 1938, 239–240.

LAVRENT'EV, M. A. The hollow charge and the principle of its action. Usp. matem. nauk, Vol. 2, No. 4 (76), 1957, 41–56.

LEVI-CIVITÀ, T. Determination rigoureuse des ondes permanentes d'ampleur finie. Math. Ann. **93**, 1925, 264–314.

LITTMAN, W. On the existence of periodic waves for velocities close to to critical. Sbornik on "Theory of surface Waves". Moscow, Izd-vo Inostr, Lit-ry, 1959.[1]

MOISEEV, N. N. On the lack of uniqueness of possible forms of steady flows of a heavy fluid for Fronde numbers close to unity. P.M.M. **21**, No. 6, 1957.

---

[1] The original of this paper could not be traced (Ed.).

MOISEEV, N. N. On the flow of a heavy fluid over a wavy bottom. P.M.M., **21**, No. 1, 1957, 15–20.

MOISEEV, N. N. and TER-KRIKOROV, A. M. On wave motions for velocities close to critical. Trudy MFTI, No. 3, 1959, 36–61.

MOISEEV, N. N. and TER-KRIKOROV, A. M. The non-linear theory of long waves. Trudy MFTI,

NEKRASOV, A. T. On waves of stationary form. Izv. Ivanovo-Voznesenskogo Politekhn. in-ta, No. 3, 1921, 52–65.

RAYLEIGH, J. W. S. On waves. Phil. Mag. (5), 1, 1876, 257–259.

RAYLEIGH, J. W. S. On periodical irrotational waves at the surface of deep water. Phil. Mag. **33**, 1917, 381–389.

SRETENSKII, L. N. On a method for the determination of waves of finite amplitude. Izv. AN SSSR Otd. Tekhn. nauk, No. 5, 1952, 688–697.

WEINSTEIN, A. Sur la vitesse de propagation de l'onde solitaire. Rend. Accad. Lincei (6), 3, 1926, 463–468.

WEINSTEIN, A. Sur un problème aux limites dans une bande indéfinie. C. R. Acad. Sci., **184**, 1927, 497–499.

III. QUASI-CONFORMAL TRANSFORMATIONS

Monographs

BERS, L. Mathematical aspects of subsonic and transonic gas dynamics. Wiley, New York, 1958.

KÜNZI, H. P. Quasikonforme Abbildungen. Springer, Berlin, 1960.

VEKUA, N. I. Generalized analytic functions, Fizmatgiz, 1959.

VOLKOVYSKII, L. I. Quasi-conformal transformations. L'vov, 1954.

Papers

AHLFORS, L. On quasi-conformal mappings, Sbornik on "Spaces of Riemann Surfaces". Moscow, Izd-vo Inostr. Lit-ry, 1961.[1]

BELINSKII, P. P. The theorem of the existence and uniqueness of quasi-conformal mappings. Usp. matem. nauk, Vol. 6, No. 2, (42), 1951, 145.

BELINSKII, P. P. On the search for quasi-conformal mappings. Dokl. AN SSSR, **91**, 1953, 99–998.

BELINSKII, P. P. On the metric properties of a quasi-conformal transformation. Dokl. AN SSSR, **93**, 1953, 589–590.

BELINSKII, P. P. On the variations of a quasi-conformal mapping. Usp. matem. nauk. Vol. 11, Bo. 5, (71), 1956, 93–95.

---

[1] The original of this paper could not be traced (Ed.),

BERS, L. Existence and uniqueness of a subsonic flow past a given profile. Comm. on Pure and Appl. Math., **7**, 1954, 441–504.

BERS, L. and NIRENBERG, L. On a representation theorem for linear elliptic systems with discontinuous coefficients and its application. Conv. Intern. sulle Equat. Deriv. e part. Roma, 1955.

BOIARSKII, B. V. Homomorphic solutions of the Beltrami system. Dokl. AN SSSR, **102**, 1955, 661–664.

BOIARSKII, B. V. The generalized solutions of first order differential equations of the elliptic type with discontinuous coefficients. Matem. sbornik **43** (85), 4, 1957, 451–503.

CHAPLYGIN, S. A. On gas jets. Sobr. Coch. Vol. II, Moscow, Gostekhizdat, 1948.

DANILIUK, I. I. Some problems of the theory of elliptic differential systems and quasi-conformal transformations. Nauch. zap. un-ta, L'vov (38), 1956, 75–89.

DANILIUK, I. I. On mappings corresponding to solutions of equations of the elliptic type. Dokl. AN SSSR, **120**, 1958, 17–20.

GERGEN, J. J. and DRESSEL, F. G. Mapping for elliptic equations. Trans. Amer. Math. Soc., **77**, 1954, 151–178.

GRÖTZSCH, H. Über die Verzerrung bei schlichten nichtkonformen Abbildungen und über eine damit zusammenhängende Erweiterung des Picardschen Satzes. Ber. Verh. Sächs. Akad. Wiss. Leipzig, **80**, 1928.

LAVRENT'EV, M. A. Sur une classe de représentations continues. Matem. Sbornik, **42**, 1935, 407–424.

LAVRENT'EV, M. A. On a class of quasi-conformal mappings and on gas jets. Dokl. AN SSSR, **20**, 1938, 343–346.

LAVRENT'EV, M. A. Quasi-conformal mappings and their derived systems. Dokl. AN SSSR, **52**, 1946, 287–290.

LAVRENT'EV, M. A. The general problem of the theory of quasi-conformal mappings of plane regions. Matem. Sbornik, **21** (63), 1947, 285–320.

LAVRENT'EV, M. A. and SHABAT, B. V. A geometric property of the solutions of non-linear systems of equations with partial derivatives. Dokl. AN SSSR, **112**, 1957, 810–811.

LAVRENT'EV, M. M. On the Cauchy problem for linear elliptic equations of the second order. Dokl. AN SSSR, **112**, 1957, 195–197.

LEHTO, O. On the differentiability of quasi-conformal mappings with prescribed complex dilatation. Ann. Acad. Sci. Fenn., Ser. AI, 275, 1960.

LEHTO, O. and VIRTANEN, K. I. On the existence of quasi-conformal mappings with prescribed complex dilatation. Ann. Acad. Sci. Fenn., Ser. AI, 274, 1960.

MEN'SHOV, D. E. Les conditions de monogénéité. Act. Sci. et Ind., 329, 1936, 1–52.

MORI, A. On quasi-conformality and pseudo-analyticity. Trans. Amer. Math. Soc., 84, 1956.

Pesin, I. N. A metric property of Q-quasi-conformal mappings. Matem. Sbornik, **40** (82), 1956, 281–294.

Polozhii, G. N. A theorem on the preservation of a region for certain elliptic systems of differential equations and its application. Matem. Sbornik, **32** (74), 1953, 485–492.

Shabat, B. V. On generalized solutions of a system of partial differential equations. Matem. Sbornik, **17** (59), 1945, 193–210.

Shabat, B. V. On mappings corresponding to solutions of Carleman systems. Usp. matem. nauk., Vol. 11, No. 3 (63), 1956, 203–206.

Shabat, B. V. The geometric meaning of the concept of ellipticity. Usp. matem. nauk., Vol. 12, No. 6 (78), 1957, 181–188.

Shabat, B. V. On mappings corresponding to solutions of strongly elliptic systems. Part of "A study of contemporary problems of the theory of functions of a complex variable." Fizmatgiz, 1960.

Shabat, B. V. On the concept of a derived system in the sense of M. A. Lavrent'ev. Dokl. AN SSSR, **136**, 1961.

Shapiro, Z. Ia. On the existence of quasi-conformal mappings. Dokl. AN SSSR, **30**, 1941, 685–687.

Vekua, N. I. Systems of first order differential equations of the elliptic type and boundary value problems with applications to shell theory. Matem. Sbornik, **31** (73), 2, 1952, 217–314.

Vekua, N. I. The problem of reduction to the canonical form of differential forms of the elliptic type and the generalization of the Cauchy-Riemann system. Dokl. AN SSSR, **100**, 1955, 197–200.

Volkovyskii, L. I. Quasi-conformal mappings and the problem of conformal patching. Ukr. matem. Zh., **3**, 1951.

INDEX

A CATALOG OF SELECTED

DOVER BOOKS

IN ALL FIELDS OF INTEREST

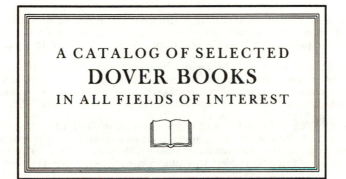

A CATALOG OF SELECTED DOVER BOOKS IN ALL FIELDS OF INTEREST

SIR HARRY HOTSPUR OF HUMBLETHWAITE, Anthony Trollope. Incisive, unconventional psychological study of a conflict between a wealthy baronet, his idealistic daughter, and their scapegrace cousin. The 1870 novel in its first inexpensive edition in years. 250pp. 5⅜ × 8½. 24953-0 Pa. $5.95

LASERS AND HOLOGRAPHY, Winston E. Kock. Sound introduction to burgeoning field, expanded (1981) for second edition. Wave patterns, coherence, lasers, diffraction, zone plates, properties of holograms, recent advances. 84 illustrations. 160pp. 5⅜ × 8¼. (Except in United Kingdom) 24041-X Pa. $3.50

INTRODUCTION TO ARTIFICIAL INTELLIGENCE: SECOND, ENLARGED EDITION, Philip C. Jackson, Jr. Comprehensive survey of artificial intelligence—the study of how machines (computers) can be made to act intelligently. Includes introductory and advanced material. Extensive notes updating the main text. 132 black-and-white illustrations. 512pp. 5⅜ × 8½. 24864-X Pa. $8.95

HISTORY OF INDIAN AND INDONESIAN ART, Ananda K. Coomaraswamy. Over 400 illustrations illuminate classic study of Indian art from earliest Harappa finds to early 20th century. Provides philosophical, religious and social insights. 304pp. 6⅝ × 9⅝. 25005-9 Pa. $8.95

THE GOLEM, Gustav Meyrink. Most famous supernatural novel in modern European literature, set in Ghetto of Old Prague around 1890. Compelling story of mystical experiences, strange transformations, profound terror. 13 black-and-white illustrations. 224pp. 5⅜ × 8½. (Available in U.S. only) 25025-3 Pa. $5.95

ARMADALE, Wilkie Collins. Third great mystery novel by the author of *The Woman in White* and *The Moonstone*. Original magazine version with 40 illustrations. 597pp. 5⅜ × 8½. 23429-0 Pa. $9.95

PICTORIAL ENCYCLOPEDIA OF HISTORIC ARCHITECTURAL PLANS, DETAILS AND ELEMENTS: With 1,880 Line Drawings of Arches, Domes, Doorways, Facades, Gables, Windows, etc., John Theodore Haneman. Sourcebook of inspiration for architects, designers, others. Bibliography. Captions. 141pp. 9 × 12. 24605-1 Pa. $6.95

BENCHLEY LOST AND FOUND, Robert Benchley. Finest humor from early 30's, about pet peeves, child psychologists, post office and others. Mostly unavailable elsewhere. 73 illustrations by Peter Arno and others. 183pp. 5⅜ × 8½. 22410-4 Pa. $3.95

ERTÉ GRAPHICS, Erté. Collection of striking color graphics: *Seasons, Alphabet, Numerals, Aces* and *Precious Stones*. 50 plates, including 4 on covers. 48pp. 9⅜ × 12¼. 23580-7 Pa. $6.95

THE JOURNAL OF HENRY D. THOREAU, edited by Bradford Torrey, F. H. Allen. Complete reprinting of 14 volumes, 1837–61, over two million words; the sourcebooks for *Walden*, etc. Definitive. All original sketches, plus 75 photographs. 1,804pp. 8½ × 12¼. 20312-3, 20313-1 Cloth., Two-vol. set $80.00

CASTLES: THEIR CONSTRUCTION AND HISTORY, Sidney Toy. Traces castle development from ancient roots. Nearly 200 photographs and drawings illustrate moats, keeps, baileys, many other features. Caernarvon, Dover Castles, Hadrian's Wall, Tower of London, dozens more. 256pp. 5⅜ × 8¼. 24898-4 Pa. $5.95

CATALOG OF DOVER BOOKS

AMERICAN CLIPPER SHIPS: 1833–1858, Octavius T. Howe & Frederick C. Matthews. Fully-illustrated, encyclopedic review of 352 clipper ships from the period of America's greatest maritime supremacy. Introduction. 109 halftones. 5 black-and-white line illustrations. Index. Total of 928pp. 5⅜ × 8½.
25115-2, 25116-0 Pa., Two-vol. set $17.90

TOWARDS A NEW ARCHITECTURE, Le Corbusier. Pioneering manifesto by great architect, near legendary founder of "International School." Technical and aesthetic theories, views on industry, economics, relation of form to function, "mass-production spirit," much more. Profusely illustrated. Unabridged translation of 13th French edition. Introduction by Frederick Etchells. 320pp. 6⅛ × 9¼. (Available in U.S. only) 25023-7 Pa. $8.95

THE BOOK OF KELLS, edited by Blanche Cirker. Inexpensive collection of 32 full-color, full-page plates from the greatest illuminated manuscript of the Middle Ages, painstakingly reproduced from rare facsimile edition. Publisher's Note. Captions. 32pp. 9⅜ × 12¼. 24345-1 Pa. $4.95

BEST SCIENCE FICTION STORIES OF H. G. WELLS, H. G. Wells. Full novel *The Invisible Man,* plus 17 short stories: "The Crystal Egg," "Aepyornis Island," "The Strange Orchid," etc. 303pp. 5⅜ × 8½. (Available in U.S. only)
21531-8 Pa. $4.95

AMERICAN SAILING SHIPS: Their Plans and History, Charles G. Davis. Photos, construction details of schooners, frigates, clippers, other sailcraft of 18th to early 20th centuries—plus entertaining discourse on design, rigging, nautical lore, much more. 137 black-and-white illustrations. 240pp. 6⅛ × 9¼.
24658-2 Pa. $5.95

ENTERTAINING MATHEMATICAL PUZZLES, Martin Gardner. Selection of author's favorite conundrums involving arithmetic, money, speed, etc., with lively commentary. Complete solutions. 112pp. 5⅜ × 8½. 25211-6 Pa. $2.95

THE WILL TO BELIEVE, HUMAN IMMORTALITY, William James. Two books bound together. Effect of irrational on logical, and arguments for human immortality. 402pp. 5⅜ × 8½. 20291-7 Pa. $7.50

THE HAUNTED MONASTERY and THE CHINESE MAZE MURDERS, Robert Van Gulik. 2 full novels by Van Gulik continue adventures of Judge Dee and his companions. An evil Taoist monastery, seemingly supernatural events; overgrown topiary maze that hides strange crimes. Set in 7th-century China. 27 illustrations. 328pp. 5⅜ × 8½. 23502-5 Pa. $5.95

CELEBRATED CASES OF JUDGE DEE (DEE GOONG AN), translated by Robert Van Gulik. Authentic 18th-century Chinese detective novel; Dee and associates solve three interlocked cases. Led to Van Gulik's own stories with same characters. Extensive introduction. 9 illustrations. 237pp. 5⅜ × 8½.
23337-5 Pa. $4.95

Prices subject to change without notice.

Available at your book dealer or write for free catalog to Dept. GI, Dover Publications, Inc., 31 East 2nd St., Mineola, N.Y. 11501. Dover publishes more than 175 books each year on science, elementary and advanced mathematics, biology, music, art, literary history, social sciences and other areas.